ATOMIC
FRAGMENTS

ATOMIC FRAGMENTS

A Daughter's Questions

MARY PALEVSKY

UNIVERSITY OF

CALIFORNIA PRESS

Berkeley   Los Angeles   London

University of California Press
Berkeley and Los Angeles, California

University of California Press, Ltd.
London, England

Library of Congress Cataloging-in-Publication Data

Palevsky, Mary, 1949–
Atomic fragments : a daughter's questions / Mary Palevsky.
   p.   cm.
   Includes bibliographical references and index.
   ISBN 0-520-22055-2 (cloth : alk. paper)
   1. Manhattan Project (U.S.)—History.   I. Title.

QC773.3.U5 p35   2000
355.8'25119'0973—dc21   99-087422
        CIP

Manufactured in the United States of America

09   08   07   06   05   04   03   02   01   00

10 9 8 7 6 5 4 3 2 1

The paper used in this publication meets the minimum
requirements of ANSI/NISO Z39.48-1992 (R 1997)
(*Permanence of Paper*).

The author is grateful for permission ro reprint excerpts
from: Freeman Dyson, "Non-use and Non-violence," talk
given at Oak Ridge Symposium of Nuclear Weapons, May
4, 1995; Bernard J. Lonergan, "Insight: A Study of Human
Understanding," The Philosophical Library, 1957.

For my parents

We are struggling to come to terms with a drama
too large for a single mind to comprehend.

FREEMAN DYSON
"Non-Use and Non-Violence"

The totality of documents cannot be interpreted scientifically
by a single interpreter or even a single generation of interpreters.
There must be a division of labour, and the
labour must be cumulative.

BERNARD LONERGAN
Insight: A Study of Human Understanding

CONTENTS

PREFACE

September 1, 1989, marked the fiftieth anniversary of Hitler's invasion of Poland. Six years later, after a series of half-century commemorations, our nation celebrated the end of World War II, while fiercely debating the atomic bombings of Hiroshima and Nagasaki that brought it to a close. This era of public remembering coincided with a private turning point in my life: it was the season of my parents' deaths, and of my mourning for them. Through my mother and father, I am connected to one of the great and controversial issues of the war—the creation of the atomic bomb.

This is a study in memory and meaning, an exploration of the intersection of the personal and public, through life spans and across generations. Questions about the moral and ethical implications of the bomb have always been in the background of my life. I write in an attempt to organize and freeze the experience of stepping into a stream that has been flowing through my subconscious for as long as I can remember. Over the years there have been times when these issues have come to the foreground. This occurred most powerfully during the fiftieth anniversary of World War II's end.

A January 1995 meeting with Nobel laureate Hans A. Bethe proved to be the starting point from which I traveled to interview sixteen Manhattan Project scientists and several of their colleagues. Our dialogues focused on the web of relationships connecting the war, the bomb, science, and society at large. The controversies surrounding the fiftieth

anniversary of the war's end served as the backdrop for many of our discussions and brought to light the ways in which questions of morality
and ethics exist in the real world of human history. The multiple narra-
tives emerging a half century after the making of the bomb formed the
warp on which the weft of the interviews was anchored and woven.

It was in the context of the discovery of fission, followed by World
War II, that traditionally academic scientists, fearing a Nazi nuclear
bomb, became the creators of the first weapon of mass destruction. My
endeavor was to comprehend the ways in which individual scientists
made choices about the bomb and made sense of their work. At best
our conversations were creative encounters during which we struggled
with the meaning of one of the twentieth century's most daunting lega-
cies. This work is an attempt to make explicit the human intersubjec-
tivity that embodies such meaning.

The location for our dialogues was "the laboratory." I am referring
not to one laboratory but to a collective of the many historical places
where the scientists have spent their working lives. There is the "Metallurgical Lab," the Manhattan Project site at the University of Chicago.
Its name was a code, meant to disguise the real activities of Enrico
Fermi and his colleagues as they worked to create the first man-made
chain reaction. Suspended in the midst of a great university in a major
American city, the shifting boundaries of the Met Lab were defined not
by architecture but by awareness. Fermi's atomic pile was hidden
beneath the West Stands of Stagg Field, the university's athletic sta-
dium. Even Robert Maynard Hutchins, the university's president, did
not know what was transpiring there.

In contrast, "Site Y," Los Alamos, was built in the isolation of the
northern New Mexico mesas, on the evacuated campus of a private
boys' school. Under J. Robert Oppenheimer's leadership, scientists
from institutions nationwide and from Europe joined intellectual forces
to produce the bomb. Many Manhattan Project scientists came from or
had studied in the great tradition of the laboratories of Europe's ancient
universities, Cambridge and Oxford, Göttingen and Tübingen; in
Copenhagen's Niels Bohr Institute; at Fermi's Roman laboratory; and
with Irène and Frédéric Joliot-Curie in Paris.

Refugee physicists Lise Meitner and Otto Robert Frisch were in the Swedish countryside when they discovered nuclear fission. Together during the 1938–39 Christmas holiday, Meitner and her nephew, Frisch, labored over experimental results that radiochemist Otto Hahn and his assistant, Fritz Strassman, had sent from the Kaiser Wilhelm Institute in Berlin. Their discussions during long walks in the snow led to the insight that Meitner's German colleagues had split the uranium nucleus. Thus the laboratory also exists in each scientist's consciousness.

The first laboratory in my experience was Brookhaven National Laboratory, where my father worked as an experimental nuclear physicist. Unlike the fine old institutes of Europe or their elegant Ivy League descendants, Brookhaven is plain. A converted World War I army barracks at Camp Upton on the Long Island flatlands is where I went with my father to meet his international colleagues and tour the giant machines they were building. Although nuclear weapons are not made at Brookhaven, other post–World War II laboratories are devoted to them. There is, of course, Los Alamos. Then, in 1952, the University of California Radiation Laboratory at Livermore was built, inland and equidistant from the halls of Stanford and the Berkeley hills. The scientists move among the many laboratories with ease.

I traveled to Cornell University and to the Institute for Advanced Study. I saw the great earthworks of the Stanford Linear Accelerator and the circular, secular cathedrals of Lawrence Berkeley Laboratory. I visited the site of the first chain reaction at the University of Chicago, where Henry Moore's bronze sculpture "Nuclear Energy" now stands. Moore hoped to evoke the feeling of a cathedral. Chicago physicist Robert Sachs explained to me that some people say the work suggests a mushroom cloud and others see a human skull. With the trinity of explosions at Alamogordo, Hiroshima, and Nagasaki, the world became the laboratory.

Throughout the interview process I kept a personal journal, recording observations, examining insights, and developing ideas and further questions. I have included selected journal entries in the text. The interviews tended to cover a broad range of subject matter; however, I have

Figure 1. "Nuclear Energy" by Henry Moore, 1967, on the site of the first controlled, self-sustained nuclear chain reaction, December 2, 1942, the University of Chicago.

focused on those themes that are central to my inquiry. As a rule, interview quotations are in chronological order, but on occasion, to clarify a particular discussion, quotations are grouped according to theme. The reader is asked to bear in mind that the quotations represent spoken, not written, words. In the interest of readability, the normal redundancies of speech have been taken out and bridge words and phrases have been inserted when needed to express language understood in conversation.

Although my dialogues with the scientists were intersubjective, there were also intrasubjective aspects to the work that have relevance on many levels: emotional, psychological, intellectual, rational, and moral. I wrote this chronicle in the "relational I"—an effort to stay true to my interiority in dialogue with others. Looking in and reaching out, I tried to bridge the gap with words.

Mary Palevsky

ACKNOWLEDGMENTS

It is with deep gratitude that I acknowledge Matthews M. Hamabata for his steadfast and insightful responses to my ideas, both subject matter and method, during my years in the doctoral program at the Fielding Institute and beyond. He has read my work with care through all of its stages, from the thesis to late drafts of the manuscript. During my graduate career, I benefited from the encouragement and critique of Elizabeth Douvan, David Rehorick, and Jody Veroff.

I thank Joan M. Bowman and Richard P. Appelbaum for their friendship, moral support, and readings of early drafts.

I am grateful to the men and women who took the time to meet with me and to engage in the dialogue during the last four years. I am indebted to Hans A. Bethe for his unfailing generosity; to Herbert F. York for his willingness to explore the many facets of my questions; and to David Hawkins for his humane spirit, expressed through meetings, letters, and long telephone calls. And I thank Rose Bethe, Frances P. Lothrop Hawkins, and Sybil York for welcoming me into their homes for meals and conversations about their recollections of the war years and their views of the events. I also benefited from our discussions of other subjects of mutual interest.

I appreciate the willingness of Edward Teller and Joseph Rotblat to meet with me and to answer additional questions during the intervening years. Patricia French of Edward Teller's office and Tom Milne of the Pugwash London office were invaluable in making arrangements

for the interviews and responding to my follow-up requests. I thank
Philip and Phylis Morrison and the late Robert Wilson and Jane Wilson, not only for the initial dialogue, but for continuing the conversations during times of illness.

Although I may not explicitly refer to our conversations, I interviewed others who contributed substantially to my understanding of
the issues addressed. My thanks to Mildred and Marvin Goldberger
and John A. Simpson for conveying to me the spirit of the Chicago project. I am grateful to Robert Sachs for our meetings and for many
months of follow-up correspondence. I also thank Harold Agnew,
Berlyn Brixner, Robert Christy, Dale Corson, Sidney D. Drell, Freeman
J. Dyson, Richard Garwin, Kurt Gottfried, Kenneth Greisen, Walter
Kohn, Boyce MacDaniel, Michael M. May, W. K. H. Panofsky, the late
Glenn T. Seaborg, and Victor and Duscha Weisskopf. Adele Marshall,
whom I had not seen since childhood, conveyed the atmosphere of
Brookhaven National Laboratory's early years. And I am most grateful
to Robert Adair for telling me his soldier's story. I learned much during
the course of these conversations, but I am, naturally, solely responsible
for what follows.

At the University of California Press, I thank Stanley Holwitz for his
willingness to consider the work of a first-time book author and for his
wise counsel. I benefited from Scott Norton's expert guidance throughout the process.

I am grateful to Janice Amaimo Bowles and Carolyn Saper for providing bed, breakfast, a personal taxi service, and so much more on my
journeys to San Diego and Chicago. I thank my family and friends for
their sustaining affection.

Joseph N. Granados helped me to create a room of my own. I labored
surrounded by his love.

Broken Vessel

I remember standing in the living room of our Long Island home when I was about ten years old, carefully examining the series of photographs that recorded the first sixty seconds of the nuclear age. They were taken in New Mexico, on July 16, 1945, at the atomic bomb test, code named Trinity. As a little girl, I wondered how the strange, silvery bubble in the initial frame could grow larger until it finally became the huge billow of smoke, ash, and dust filling the sky in the final ones. In my child's mind, the mushroom cloud was "the bomb." I did not understand that the first, beautiful, iridescent dome rising from the flat New Mexican desert was the deadly, expanding fireball.

My mother, who had worked in the Los Alamos optics group, brought the pictures home with her after the war. My parents, both Chicagoans, met, fell in love, and married while working on the creation of the atomic bomb. They were young scientific workers, first at Chicago's Metallurgical Laboratory and later at the bomb-building lab in New Mexico. Their routes to the Manhattan Project labs were like those of countless scientists of their generation. They were not among the elite handpicked by the Metallurgical Lab leader, Arthur Holly Compton, or by the Los Alamos director, Robert Oppenheimer.

My father, (Harry Palevsky,) received a scholarship to Northwestern University to study electrical engineering. His parents were poor, and a family crisis ensued when they could not afford to buy him a slide rule. Fortunately, they eventually obtained one, and he was able to proceed

Eastern European Jew.

Figure 2.

with his studies and earn a bachelor's degree. A Northwestern professor recommended him for his first war-related work, as a
civilian employee of the Naval Ordnance Laboratory near Washington,
D.C., where he developed mine detection equipment under the future
two-time Nobel laureate, John Bardeen. Then he went to Chicago's
Metallurgical Laboratory, where he worked on instrumentation to
detect radiation. It was there that Enrico Fermi and his colleagues had

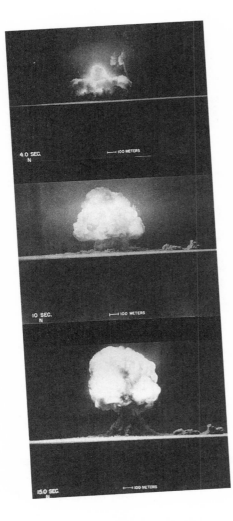

Figure 3.

achieved the first man-made, controlled nuclear chain reaction on December 2, 1942.

My mother, Elaine Sammel, attended Wright Junior College in Chicago and then worked as a dancer with the USO. At the end of the first tour, keeping a promise to her dying father, she completed her undergraduate degree in physics at the University of Chicago. It was through a Chicago professor that she got her job in the Met Lab's optics group. My

*Figure 4. My mother on
tour with the USO, 1943.*

parents met at a party and began dating. A short time later, my father
responded to a call for more technical workers in Los Alamos, where the
atomic bomb was actually being built. He was assigned to the electronics
group and soon learned good people were needed in optics. He recommended my mother, who joined him in Los Alamos. In early July 1945
she wrote a joyful letter home, telling her widowed mother that my
father had proposed marriage and she had accepted. After asking for her
mother's blessing, she cautioned, "The actual time and place of our marriage depend too much on too many things which I cannot tell you
about. But these things can be discussed later." My parents bought my
mother's wedding band from a Pueblo jeweler in Santa Fe and were
married there on July 25, 1945, nine days after the Trinity bomb test. My
mother was twenty-two years old, my father twenty-five.

Within three weeks of their wedding, the atomic bombs were
dropped on Hiroshima and Nagasaki, and the war ended. When my
father returned to Chicago, he confided in his sister, Helen, that from
the moment he heard of the bombings of the Japanese cities, he had

thought they were wrong. And he told her he would never work on weapons again. He completed his education and went on to a long career as an experimental nuclear physicist at Brookhaven National Laboratory on Long Island. Like many of the women who worked on the Manhattan Project, my mother did not pursue a career in science. She returned to her first love, dance, and owned a studio in our community for many years. Most of the girls in my town and of my generation studied ballet with Mrs. Palevsky.[1]

In 1981, after working at Brookhaven for more than thirty years, my father was forced to take an early retirement. He was only sixty-one years old but had suffered the first in a series of small strokes. That summer I visited my parents on Long Island during the Fourth of July holiday. After a few days, I understood with shocking certainty that my father had changed, that the man I had known and loved all my life would never be the same. The vessel of his mind was breaking. I wrote in my journal:

July 5, 1981
Brookhaven

My father is like a Christmas ornament that has been dropped and not broken—but when I look closely, I can see that tiny fault lines are etched in the surface, threatening to break open at any moment. My heart aches as I cradle this delicate, shattered life in my hands.

When my father retired, his doctors said that he would probably live for five years. Perhaps they could not see the fierce stubbornness behind his increasingly passive demeanor. I returned home to California and wept deeply. It was the beginning of a long mourning that would continue beyond his death nine years later. During that time, he weakened physically and mentally and suffered severe pain in one of his feet because of diminished circulation to his legs. Once a generous and gregarious man with a contemplative side, my father withdrew, was often remote—even cold. Before he was seventy he possessed the behavior and demeanor of a much older man, and my

Figure 5. My father, Los Alamos, 1945.

mother, silently grieving at every step, was reluctantly transformed into his caretaker.

During Christmas 1987 I visited my parents in their New York City apartment. By this time they had sold our family home in Brookhaven. My mother told me that although she loved the house, she could no longer maintain it by herself, without my father's help. But she failed to reveal her own secret: her body was already being weakened by the cancer that would soon take her life.

I remember a conversation my mother and I had, almost in passing, during that holiday visit. We were probably cleaning, or preparing dinner. Earlier in the day, some of my childhood friends had stopped by for a visit. I had read in their faces the shock at my father's premature aging; however, they had been too polite to say anything. My friends had left, and my mother stood before me declaring that she thought someone should write about my father's life and times. "He has seen and done such interesting things," she said, "and he remembers them." Her eyes filled with tears, and she angrily spit out the words, "Now all anyone sees when they look at him is a doddering old man. They don't realize that he's still in there." When I asked if she was going to write the book, she shook her head no and walked back to the kitchen.

One year later the family gathered around my mother for her final

Christmas. Looking back, I see that we all knew then that she would not survive the cancer that was ravaging her body, traveling from breast to lungs to brain. But as so many do, we hoped against hope—the deterioration of this bright and vital woman had come too quickly and unexpectedly. My mother lived fully until the end. During the last years of her life, she had taken up the study of dispute resolution, becoming a valued member of the staff at the Queens Mediation Center in New York. Her love of peacemaking was most profoundly expressed when, just a few weeks before her death, she gave a speech about mediation in international affairs for the National Council of Women of the United Nations.

They say that the dying experience a life review. However, I vividly remember entering my mother's hospital room and seeing my own choices pass before my eyes. At the age of thirty-nine I had made my plans for the future and had figured out what the rest of my life would look like. With many years of experience in business and an ability with numbers, I had recently enrolled in courses to become a certified public accountant. But in the presence of my sleeping mother, I felt that my sensible, well-reasoned decision had been made from a deep resignation. Witnessing the woman who had birthed me encounter her end-of-life crisis, I felt cowardly. I had only the vaguest sense that there must be something more to my life's work. The least I could do was risk discovering whether my imaginings were real.

Later that day, during a private moment, when my mother and I admitted to each other what we could not face within ourselves, she asked, "What will become of your father?" I answered her the only way I knew how. "Don't worry, Mom, we'll take care of him." She pleaded, "But he needs me," and I replied, "I know he needs you, Mom, but we'll do the best we can." Thus I made a promise to my mother at our moment of love at last sight.*

My mother was dead before the New Year at age sixty-six. We brought my father to California, where my husband, Joseph, and I cared for him until his death two years later. Several months before he died,

---

* A nod to Walter Benjamin.

Figure 6. My mother six months before her death, 1988.

remembering my conversations with my mother, I asked my father if he would like to record his memoirs. He wanted to do this, so we spent several afternoons in my sunny California living room as he recounted the essential moments of his life. Speaking slowly and carefully into the microphone, he recalled growing up in Chicago, the first-born son of poor Eastern European Jews. His mother had come from a family of rabbis, and he told me stories that embodied her moral lessons.

My father remembered teaching himself how radios work and, as a teenager, becoming famous in his neighborhood for fixing them. And he recalled his pursuit of a career in nuclear physics with its thrilling beginnings at the Manhattan Project's Metallurgical Lab. He told me of being among the young scientists surrounding Leo Szilard as he held court. The Hungarian-born physicist's ideas, my father said, had influenced his own developing social conscience. He described the genius and generosity of Enrico Fermi and smiled as he recalled being at Los Alamos and first encountering Richard Feynman's penetrating insight into physical reality.

Although he felt privileged to have worked alongside some of the century's greatest scientists to end the war, my father was deeply troubled by the terror they had wrought to achieve the peace. I had long known his profound misgivings about the use of the bomb and his complicated feelings about his participation in its creation. As we spoke, I had the sense that his effort to reconcile the moral complexities of the bomb was being transmitted to me. When he died, it became his legacy.

Traveling with my parents to the ends of their lives constituted a major turning point in my own. I never returned to my accounting classes. Eighteen months after my father's death, I entered graduate school. I could not know then that I would study the moral legacy of the bomb. I only knew that I wanted to continue an education that I had left behind in my confusion amid the societal explosion we now label "the sixties." Early in my graduate career, during a human development seminar, a colleague raised the issue of the atomic bomb. She characterized as amoral a Manhattan Project scientist she knew who felt no remorse for the fates of Hiroshima and Nagasaki. I understood why she made that judgment, but for me the question was not at all simple or clear. I replied that this man's reasoning and his support of the use of the bomb were not, in and of themselves, proof of his amorality. I suggested that for him using the bombs to end the war may have been moral. Naturally, our conversation brought to mind my own relationship with the bomb.

A year and a half later, during the commemoration of the fiftieth anniversary of D day, for the first time in my life I became completely engrossed in the news stories about World War II and those who had fought it. Watching television documentaries, I was struck by how little I knew about the Normandy invasion and was surprised to learn that many D day veterans had never discussed their experiences with their families, in particular, their children. Listening to them reminisce, I began to sense the hell that had been World War II.

Soon I was reading the early press reports about the Smithsonian Institution National Air and Space Museum's exhibit for the fifty-year commemoration of the war's end. I heard the first public rumblings of what was to erupt into a battle over the competing meanings of the bomb and how it should be remembered. It was then that I decided to play the tapes of my father's memoirs, which had remained untouched on my bookshelf since his death. I removed the first cassette from its dusty box and listened, expecting the sound of my dead father's voice to be eerie, even frightening. Instead I found strange comfort in his halting but determined effort to speak to me through all that had separated us.

I gathered everything my father and I had done together—the three audiotaped interviews; some notes from a memoirs writing class we had taken at the local senior center; a scant printout of a family history program I tried to help him complete. I discovered a scrap of paper with my father's words scribbled on it; listening to the tapes, I heard stories without endings; I sifted through a damp cardboard box filled with files and photographs salvaged from my dad's long-abandoned Brookhaven Lab desk.

I awoke from dreams with deep unease; my memories faded and then seemed to change. Time had shifted when my mother died. I recalled that evening, when my father was reminiscing and mistook me for his sister, Helen. Perhaps his stories had carried him back in time so that he could not imagine that the dark-eyed, listening woman was his own daughter. He asked impatiently, "Don't you remember when we . . . ?" And I, leaning across my kitchen table, gently answered, "Dad, you must be thinking of Helen. I am Mary." He blinked and with a lost, then hurt look replied, "Right." This was not the book of my father's life my mother had envisioned, but then her own story had not turned out as planned.

Transcribing the tapes, I heard my father's quiet words: "When the word got out they were going to use the bomb, there were people at Chicago who opposed it. And they visited Los Alamos and talked about their opposition. They thought we ought to tell the Japanese that we were going to use it, and then, after we used it, they thought the Japanese would surrender." I stopped typing and gazed back, beyond my dad's end-of-life recollection, to the time I first remember hearing a rendition of the story. It was the 1950s, and my father, my younger brother, and I were at Brookhaven. I see this childhood picture:

Dad, Alan, and I are sitting at one of the long cafeteria tables. There is the

Figure 7. Me at age six, 1955.

noise of the lunch crowd in the background. We have just
come from seeing the giant machine where he does his experiments. We have
our food and we are excited to be with Dad at his work. Then he tells us about
the bomb, that he and Mom worked on this very important project. He says
he helped build the trigger mechanism that made the bomb explode. We ask if
that was like a trigger on a gun, and he explains it was not, that it was elec-
tronic. Then he tells us the most important thing: he and Mom didn't want
the bomb to be used on people in the war. They wanted it to be exploded on an
island in the middle of the ocean, where no one lived, but where the Japanese
people could have seen how big it was. Then they would have surrendered.
The war would have ended and no Japanese grown-ups or children would
have been hurt by the bomb my parents had helped make.

As a child, listening to this story given in love, I did not visualize the
terrible destruction of Hiroshima and Nagasaki. I could not imagine
what had actually occurred. My mind created a picture of what my
mother and father had wanted to have happen: the big bomb exploding
on an island where no one lived; families kept from harm. I did not
think of the Japanese as enemies to be feared. I envisioned adults and
children living safely and happily on the other side of the world. My
father told us that the bomb demonstrated in this way would have been
a noble effort to end the war without hurting anyone, so this is what I

saw. The story defined my parents' connection to the bomb and their personal goodness.

Thus, with the approach of the fiftieth anniversary of the atomic bomb, I began writing from a lifetime of fragments that are my father's story of his wartime work on the Manhattan Project. Now the task seemed even more urgent. What had it been like for my father and mother, and for the others who were brought together by the forces of history, chance, and talent to build the first atomic bomb? How did my parents' lives relate to the larger picture? Sometimes, like a jigsaw puzzle, aha! the pieces would fit together perfectly. Other times, like shards from long-buried vessels, the edges were worn down, the shapes changed, the colors muted. I could not possibly know whether I had put them together as they once were. Then I relied on my own imagination and judgment, my particular sense of form. One moment everything seemed to achieve a kind of unity. At others I stepped between the pieces into emptiness and was shattered. What had made sense lost meaning, my wholeness dissolved, I walked without bearings. Yet my desire remained. As I began my exploration, I discovered I was not alone in my search for the meaning in the remnants left by the bomb.

I watched as the controversy surrounding the Smithsonian's exhibit developed. The comprehensive presentation was to display the fuselage of the *Enola Gay*, the B-29 that dropped the bomb over Hiroshima. It was also to include life-sized photographs of the bomb's victims, along with artifacts from the bomb site. Opponents argued that early drafts of the script characterized the Allies' Pacific war as vengeful and racist. Veterans' groups complained that the "revisionist" exhibit lacked balance by not also addressing Japanese atrocities in China, the Bataan death march, and the prisoner of war camps.

In addition, the presentation was to explore how the bombs thrust the world into the nuclear age. Arguing against assertions that the American public had never "come to terms," critics said that it understood very well what the atomic bombings had meant: the end to the worst war in history and the saving of hundreds of thousands of American lives that would have been lost in an invasion. Members of Congress asserted that the function of such an exhibit was to honor Americans who had died fighting Nazism and Japanese militarism. At the other end of the spectrum were those who felt that the planned exhibit did not go far enough in exposing the impact of nuclear denial and secrecy on Americans, physically and psychologically, or the larger worldwide environmental and health consequences. In the end, the National Air and Space Museum displayed only the *Enola Gay*'s fuselage, along with a video of the flight crew's recollections.

I wanted to enter the larger debate from the perspective of my relationship with my father. I was particularly interested in the moral dilemmas that the scientists faced and the context in which the difficult wartime decisions were made. I also wondered how, in light of fifty years' experience, Manhattan Project participants reflected on those extraordinary times. I began by doing library research. Since my field of graduate study was human development, I had spent many hours in the humanities sections of the university library. But when I stood at the double doors marked by the sign Sciences and Engineering Library I was anxious. Crossing the threshold, I felt confused and disoriented. I was only a few feet from the stacks of sociology and philosophy texts that I explored with ease and enjoyment, but here I felt like an interloper—a spy in the house of science.*

However, driven by my desire to better understand my late parents and the forces that had shaped their lives, I wandered deeper into the stacks. As I came upon books that could hold answers to my questions, my anxiety was replaced by the excitement of discovery. I pulled the atomic bomb histories from the shelves and hungrily searched for clues in the indexes of the scientists' biographies. Bibliographic leads sent me down unexpected paths and into unanticipated corners.

During my first hours at the library, I found two books that have become treasures to me. *Leo Szilard: His Version of the Facts* contains the émigré Manhattan Project physicist's selected recollections and correspondence from 1930 through 1945. The title came from one of Szilard's recollections:

* A nod to Anaïs Nin.

In 1943 Hans Bethe from Cornell visited in Chicago and we discussed the work conducted there under the Manhattan Project in which I was involved. The things that were done and even more the things that were left undone disturbed me very much particularly because I thought (quite wrongly as we now know) that the Germans were ahead of us. "Bethe," I said, "I am going to write down all that is going on these days in the project. I am just going to write down the facts—not for anyone to read, just for God." "Don't you think God knows the facts?" Bethe asked. "Maybe he does," I said— "but not *this* version of the facts."

I was interested to read about the change in atmosphere at Chicago's Met Lab once Germany's defeat was assured and the lab's most vital work completed. Szilard recalled that he and other scientists "began to think about the wisdom of testing bombs and using bombs." "Initially," he wrote, "we were strongly motivated to produce the bomb because we feared that the Germans would get ahead of us, and the only way to prevent them from dropping bombs on us was to have bombs in readiness ourselves. But now, with the war won, it was not clear what we were working for."[2]

I also read Szilard's July 1945 petition to President Harry Truman, by which he hoped to enable the signers to go on record with their opposition, on moral grounds, to the use of the bomb against the Japanese at that stage in the war. The petition asked that, before any atomic bombings, the Japanese be informed of the terms to be imposed after the war and be given a chance to surrender.

Alice Kimball Smith's history of the atomic scientists' movement, *A Peril and a Hope*, describes the wartime evolution of an uneasy awareness among atomic scientists like my father regarding the larger social and political consequences of their work. I read a 1944 Met Lab document titled "Prospectus on Nucleonics," known as the Jeffries Report, which addressed "the dilemma of technological progress in a static world order" and warned that "technological advances without moral development are catastrophic." Smith's volume also contains a copy of the June 1945 Franck Report, which exposed the roots of my father's views on the bomb.

I had heard of Chicago's Franck Report but had only a vague sense of its contents. It is a fascinating historical document for several reasons, not the least of which is the way it addressed the relationship between science and war. A committee of Met Lab scientists, headed by the highly respected German-born Nobel laureate James Franck, had been charged with studying the social and political implications of nuclear weapons. Franck was deeply committed to the report's conclusions and recommendations. In early June 1945 he traveled to Washington, hoping to deliver it to Secretary of War Henry L. Stimson. Unable to arrange a meeting, Franck had to settle for leaving the document with Stimson's assistant. The scientist-authors predicted the coming arms race with remarkable accuracy. And most important, they linked the postwar implications of the bomb with its wartime use.

> *We cannot hope to avoid a nuclear armament race either by keeping secret from the competing nations the basic scientific facts of nuclear power or by cornering the raw materials required for such a race. . . .*
>
> *From this point of view, a demonstration of the new weapon might best be made, before the eyes of representatives of all the United Nations, on the desert or a barren island.* The best possible atmosphere for the achievement of an international agreement could be achieved if America could say to the world, "You see what sort of a weapon we had but did not use. We are ready to renounce its use in the future if other nations join us in this renunciation and agree to the establishment of an efficient international control."
>
> After such a demonstration the weapon might perhaps be used against Japan if the sanction of the United Nations (and of public opinion at home) were obtained, perhaps after a preliminary ultimatum to Japan to surrender or at least to evacuate certain regions as an alternative to their total destruction.[3]

As I read the words "a demonstration . . . on the desert or a barren island," I felt a deep sense of relief. Finally, I might begin to understand where my father's story had come from. Then, as I glanced at the report's seven signatories, one name jumped out at me, Donald J. Hughes. Don had been my father's closest friend. They had first met while working at the Naval Ordnance Lab, where, one day, Don received a call from Metallurgical Lab director A. H. Compton, under whom he had studied at Chicago, calling him back. It was Don who first told my father the secret of the chain reaction and the plans to build a bomb. And when my father returned to Chicago to visit his parents, Don recommended him for a job at the Met Lab.

After the war Don went to Brookhaven Lab, and in 1950 he invited my father to join the neutron physics group he was forming there. Brookhaven's early years were full of excitement and challenge for the young scientists, and Don and my dad grew even closer. Then, in 1960, Don died suddenly and tragically. But my father did not speak to me about his friend until thirty years later, when, as we recorded his memoirs, his own death was approaching. That Donald J. Hughes had signed the Franck Report was one more piece in the puzzle that is my attempt to understand my father.

I searched my mind for someone who might know more about Don Hughes, the Franck Report, and what the atmosphere had been like at Los Alamos during the spring and summer of 1945. Then I remembered William A. Higinbotham, a Brookhaven scientist whom I had also known from childhood. Willy had been my father's group leader at the Manhattan Project's Los Alamos laboratory. After the war they were active in the atomic scientists' movement as members of the Federation of American Scientists, of which Willy was a founder and guiding light. During the 1950s the Hughes, Higinbotham, and Palevsky families all lived on the same Long Island country road. When my father retired, Willy wrote a letter remembering his scientific work. However, he stated that most important to him had been their thirty-seven-year association at Los Alamos and Brookhaven, "as regards our proper role as scientists toward society."

I had not seen Willy for many years and remembered him as a small, sparkling man, famous for playing a mean accordion. I telephoned him, and the moment he answered, I knew from his weak voice that he was very ill. As I explained my call, he told me it was difficult for him to concentrate his thoughts but that he would be happy to respond to my questions by letter. "Sure, baby," Willy said. "I've got a lot to tell you and a lot to ask you."

I immediately wrote, sending a list of questions. I wanted to know what the young Los Alamos scientists had talked about among themselves during the months leading up to the bombings. Did they, like the Chicago scientists, discuss alternatives such as a demonstration? Were there any means for them to express opposition to the use of the bombs on civilians? Did he remember discussing such questions with my father? I do not know whether Willy Higinbotham ever saw my letter. The following month, I read his obituary in the *New York Times*. I experienced a kind of despair, having lost the chance to speak to someone who knew my father's world in a way that had always been closed to me. Another strand connecting me to my past, and to my parents, was severed. I clipped the obituary—one more fragment for my files.

In December 1994, several weeks after Willy Higinbotham's death, I woke up early thinking, why not call Hans Bethe? I knew that the German-born Nobel laureate had headed the theory division at wartime Los Alamos and had been at Cornell for nearly sixty years. Then in his late eighties, he remained active in both science and arms control. Over the years, I had read his reasoned discussions of both. If anyone could provide insight into the bomb and its era, it was Bethe. To someone of my background, the physicist was scientific royalty, so before I had time to lose my nerve I hopped out of bed, threw on my robe, splashed water on my face, and placed the call.

I did not know then that Hans Bethe would become my Janus. Facing both beginnings and endings, he stood guardian at the portal as I embarked on my quest to understand the people and times that had created the first weapon capable of breaking the vessel of the world.

CHAPTER I

Hans A. Bethe, Tough Dove

Professor Bethe answered the telephone in a deep "Hello," and I started talking. When I introduced myself and explained that my dad had been an experimental physicist at Brookhaven, he replied that although they had never met, they had corresponded. I assumed they had not known each other at Los Alamos. My father had been too far down the hierarchical ladder from Bethe. I requested an interview, and Bethe cordially consented, so we made an appointment to meet the following month at the California Institute of Technology, where he goes each winter to do astrophysics research.

Born in Germany in 1906, Bethe experienced firsthand the rise of nationalism in response to the harsh provisions of the Treaty of Versailles following World War I. In the early 1930s the talented young physicist lost his assistant professorship at the University of Tübingen. Bethe's mother was a Jew, and the first Nazi racial laws prohibited him from working as a civil servant in the state-run university. He was one of the many academics, artists, and intellectuals who fled continental Europe as fascism spread. In 1933 he took a temporary position in Great Britain and in 1935 accepted an acting assistant professorship at Cornell University. Then, in 1943, J. Robert Oppenheimer, director of the Los Alamos lab, asked Bethe to head the Theoretical Division of the bomb-building project. After the war he returned to academic life at Cornell University. He was a cofounder of the Federation of American Scientists, which lobbied for scientific openness and international control of atomic energy. The federation was partly responsible for atomic energy being under the aegis of the Atomic Energy Commission (AEC), a civilian organization, rather than the military.

Later Bethe opposed the United States' rush to make the hydrogen bomb and the development of antiballistic missiles. He served on presidential scientific advisory committees and has been a longtime supporter of nuclear test bans. An advocate of the nuclear freeze, during the Reagan years he spoke out against the development of the Strategic Defense Initiative (Star Wars). In 1967 Bethe was awarded the Nobel Prize in physics for his work on nuclear reactions, in particular, for discovering the source of energy in stars—what makes stars shine. He is now professor emeritus at Cornell, where he continues to work daily.

In January 1995 my husband, Joseph, and I made the two-hour trip from our home to Caltech for what was to be the first of my meetings with Bethe. Joseph drove, chauffeur style, while I sat in the backseat, my briefcase open and papers strewn around me. This was my initial interview outside the family circle, and I wanted to be well prepared. We met in Bethe's small, cramped office in Caltech's Kellogg lab building. He shared the room with Gerald Brown, a physicist from the State University of New York at Stony Brook, with whom he conducts research. I do not know why, but I expected that Caltech would house the eminent professor in more luxurious quarters. As it was, the room contained two old desks, some worn chairs, bookshelves, the ubiquitous blackboard covered with white chalk calculations, and a telephone that the scientists answered themselves.

The only pictures I had seen of Bethe were of his much younger days, so at first I did not discern the theoretical physicist in the slender, pale, white-haired gentleman who greeted us. However, I recognized him when his face opened in an unmistakable broad smile. We took our seats, and I asked for permission to tape the interview. I told him a little about myself and, as a kind of introduction, nervously handed him some papers about my dad. But he waved them away, saying he did not need them, he knew who my father was.

Any notion of frailty vanished when Bethe spoke; his German-accented voice was strong and unwavering. As I asked my questions, his

Figure 8. Hans Bethe and son,
Los Alamos, 1945.

eyes focused hawklike on me. He usually paused after
each one, considering his response before answering. It was immediately clear to me that Bethe possessed a powerful memory. As I queried
him about scientific history and the Allied and Nazi bomb projects, he
answered precisely, supplying names and dates to support his statements, often citing published references.

I was curious about the discovery of fission. I knew that in summer
1938, Austrian-born physicist Lise Meitner had escaped from Germany
to Holland. She and her close friend and colleague Otto Hahn continued their study of the atomic nucleus at the Kaiser Wilhelm Institute in
Berlin. That December Hahn and his assistant, Fritz Strassman, bom-
barded uranium with neutrons and produced the element barium. Uranium was then the heaviest element known, and barium is just about
half its weight. Hahn corresponded with Meitner in Sweden, where she
had settled, asking her to suggest an explanation for his "barium fantasies." Meitner, along with Otto Frisch, discovered theoretically what
Hahn and Strassman had found experimentally: the barium was a fragment of a hitherto unknown process. The uranium nucleus was, in fact,
splitting, releasing energy, and setting free at least two neutrons. Soon
thereafter Frisch conducted experiments that provided convincing

proof of the reaction and, using a term that describes the division of living cells, named the process "fission."[1]

I asked Bethe if the discovery of the fissioning nucleus—something we now take for granted—was in fact so revolutionary in the late 1930s. He answered emphatically,

> Yes, it was totally revolutionary. We knew beforehand that nuclei can change by one or two units of atomic number—artificial radioactivity— but we never had seen any reaction in which the nucleus just splits in two—completely new. Now, it had been predicted theoretically by a German woman named Ida Noddack, but nobody would believe her.

I wondered if Bethe had ever thought about what might have occurred had the discovery of fission not coincided with the prelude of the war with Germany—if there had not been a push for an atomic weapon. He responded,

> That's a very important and interesting question. Now, almost immediately, Szilard said this [fission] gives a chain reaction. Almost immediately, in January of '39, he realized that this might give an explosion. And at the same time it was clear that you could use that tremendous energy for peaceful purposes—for energy. So, I imagine that if there had been no war, people would have gone for the peaceful uses, including especially energy.

I was under the impression, from something I had read, that Bethe himself had thought the chain reaction might not be possible. When I asked if this was, in fact, the case, he corrected me.

> That is not true. There is something close to the truth. Several of my friends, including Fermi, Szilard, Teller, [physicist Eugene] Wigner, were fascinated by the chain reaction as an explosive. So I was doubtful whether that was possible, because that would require the chain reaction with fast neutrons. And for that you needed separated [U-] 235 and that was obviously very difficult. So difficult that the Germans decided right in the beginning of the war, that they couldn't do it. And by great good luck, the Germans also didn't find out that you can make a slow chain reaction with ordinary uranium with graphite. That was our great

ace in the hole. Because Fermi studying it very, very carefully showed, I guess already in '41, that this would be possible. But you needed extremely pure graphite. That is where Szilard came in so importantly.

Bethe went on to explain that after Fermi got a negative result with graphite, he suspected the trouble might be an impurity, but he had no idea what element that could be. Fortunately, Szilard had been a chemical engineer and knew that boron carbide was often used in the manufacture of graphite. And boron, it turns out, is a tremendous absorber of neutrons. Thus it would absorb the neutrons needed for a chain reaction. Szilard found a manufacturer who would produce graphite free of boron for the experiment. Bethe told me, "The Germans had a good nuclear physicist by the name of [Walther] Bothe, and he did the same experiment as Fermi—I don't know if it was exactly the same—but anyway, he found that graphite wouldn't do. He had no Szilard."[2]

During the first half of the interview, with precision and wit, Bethe told me fascinating stories about those early days of scientific discovery, leading up to the creation of the atomic weapon. But today, looking back, it seems I had really sought him out with one question. I finally mustered the courage to ask it: Were his feelings about the bomb anything like my father's? He listened with a sympathetic expression as I described my father's lifelong misgivings about the bomb. I had come to believe that the United States should have tried to bring the war to an end through diplomacy regarding Japan's retention of the imperial institution, or a demonstration of the bomb. I asked Bethe if he regretted the use of the bomb, and I was shocked by his emphatic response: "I do not."

This was my first indication of the distinction Bethe makes between the bomb's use during the war and his devotion to preventing its use after the war. I had assumed that as a thoughtful and passionate advocate of arms control, he would at least express some doubts, some second thoughts about the first use of the atomic weapon. My immediate impulse was to discount his views—but quickly following on this was my hope to understand more. I knew that Bethe had considered the question for fifty years.

After returning home, I wrote to Bethe requesting a follow-up interview to discuss his ideas about the use of the bomb in more detail. He
consented, saying that he also wanted to address arms control. I was
planning to visit my family in Maine and arranged to meet with Bethe
en route. I stopped in Manhattan to see some friends and took a plane
to Ithaca on a cold Saturday morning.

On my arrival at the hotel, I telephoned the Bethe home and arranged
a time for Mrs. Bethe to drive over to get me. I waited for her in front of
the hotel, enjoying the icy air I so miss in California, and at the
appointed hour she arrived. I had read several historical accounts of
wartime Los Alamos describing her beauty. When we met I was immediately struck by her finely chiseled features in lovely proportion. However, this is not what made Rose Bethe, then in her late seventies,
impressive to me. I quickly came to understand that she is a woman of
penetrating intelligence. And her straightforward, no-nonsense manner
means that I always know where I stand with her.

At that time the Bethes were living in the house they had shared
since the end of the war. When I arrived Professor Bethe greeted me
and we sat down in the living room, where he methodically set out the
logic of his argument in support of the atomic bombings of Japan. In
front of him on the table were two handwritten sheets of paper, one
titled "Use of the Bomb" and the other "Post-War." On each Bethe had
carefully listed his argument's main points.

He began by telling me that in his opinion three alternatives
existed for ending the war in the Pacific: blockade, invasion, or the
bomb. He did not believe surrender would have occurred independent of any of these actions but thought that Japan's peace overtures
to Moscow were doomed from the outset by Stalin's expansionist
designs on East Asia. First he discussed the blockade: "We had complete superiority in the air and at sea. The Japanese were dependent
on the import of oil; in fact, that was probably the reason why they
attacked Pearl Harbor in the first place. And without oil they couldn't
pursue the war."

A prolonged blockade, he said, would have meant starving the Japanese people. One argument against a blockade from the American point

*Figure 9. Rose Ewald (Bethe),
New Hampshire, 1938.*

of view was that the troops were eager to come home.

After surrender, even more extensive support of Japan would have been necessary. However, Bethe's most important argument against blockade was its potential impact on the Japanese populace.

> The blockade would surely have been successful, but it would have left great resentment in Japan, just as it did in Germany after the First World War. It probably would have caused the same resentment I experienced in the early twenties. The German slogan "Im Felde unbesiegt"—Never defeated in the field—was the seed for the Nazi movement. I believe that Roosevelt and likewise the British had this very much in mind when they asked for unconditional surrender. They did not want this kind of talk in either Japan or Germany.

Bethe's illustrious career as an American physicist has taken him far from the Germany of his youth. Yet his personal experience of the aftermath of World War I and the Nazi terror it nurtured caused him to seriously consider the potential danger of a prolonged blockade of Japan. This is also an important element of his argument against the second alternative, an invasion of the Japanese home islands. "There has been a lot of writing recently that an invasion would have been far less costly than was once stated. Truman's and Stimson's estimates were half a million to one million American casualties. I think these statements are likely to be correct."

I asked what he made of the discrepancies in invasion casualty estimates that were being reported in the press. Some historians argued that actual wartime estimates were much lower than Truman, Churchill, and Stimson claimed after the war. Bethe replied, "I am convinced that the low estimates of casualties which are now being bandied around are fantasy—after all, we had the experience of the islands [Iwo Jima and Okinawa]. The Japanese would now defend the homeland and the fighting would probably have been even more fierce."

I was asking about Allied casualties, but Bethe continued with a discussion of potential Japanese losses. He pointed out that in an invasion Japan's military casualties would have been higher than the Allies' and noncombatant deaths would have been tremendous. I had read that at least seventy-five thousand civilians died on Okinawa alone.

> If there had been a blockade or invasion, the firebombing would have continued. The casualties [from incendiary bombs] in Tokyo were close to those of Nagasaki. We must visualize that there would have been such bombing raids week after week. The casualties and the destruction of these would have been many times those of Hiroshima and Nagasaki.
>
> Then it is important to talk about the Soviet participation. In Yalta, in February 1945, we begged the Soviets to enter the war in the East. They did so, but only after Hiroshima. I think it was a mistake to ask them, because already at that time, February 1945, it was clear that we would have the uranium 235 bomb. There was no question that it would work. It was only a question of delivery of material.
>
> [The Soviets] *wanted* to be in, and of course conquered Manchuria immediately. Stalin wanted to occupy the northern half of Hokkaido, which is in the home islands—about one-tenth of the [Japanese] population. We would have had the same difficulties there as we had with the Soviet occupation in Germany. It would have been a very difficult situation for the Japanese.

With hindsight, Bethe considered the potential problems with the Soviets to be further proof that the decision to use the bomb was correct. However, he did not agree with those who claim that the bombings had less to do with defeating Japan and more to do with postwar relations with the Soviet Union. The bomb, he asserted, was used to end the war quickly and save the lives of Allied soldiers. As he spoke, I recalled another element of my father's mixed feelings: his own younger brother, who had been stationed in the Pacific, came home alive. Would he have survived without the bomb?

We turned to the question of why the bomb had been developed in the first place. Many émigré scientists, who until then had lived in the great ivory tower of European scientific tradition, joined the bomb project because they feared a Reich made invincible by a Nazi nuclear weapon. I asked Bethe what he recalled of that era.

> We all went to war against Nazi Germany, which we felt was a menace to the world. After all, the Nazis had taken continental Europe. I remember their slogan "Today we own Germany, tomorrow the whole world." The refugees from Europe [and] most of the leading American scientists worked with great enthusiasm on something which might win the war against Germany.

When I asked Bethe why Germany's defeat did not change the reason for developing the bomb, he replied that from the outset he considered the use of the bomb a foregone conclusion. The goal of the Manhattan Project was to build an atomic weapon. The war had been long and hard. When built, the bomb would be used, on Germany or on Japan. I wondered what he thought of the idea of a demonstration and of the Franck committee's recommendations. He told me he never considered the Franck Report a viable plan.

> The demonstration would not have been effective. It probably would have been seen by some [Japanese] military. It was quite unlikely that they would have told the emperor, and the emperor was the key to surrender. Without him, the war would have lasted much longer. The Japanese military were fanatical, the cabinet had three military and three civilian ministers. Even after Hiroshima [the] three civilians were for surrender, while [the] three military were still for carrying on. The emperor decided, "We surrender. I cannot see that kind of destruction go on,

one city after another." It seems to me that the war could not have
been ended by a demonstration.[3]

So Bethe came to the third alternative, the use of the nuclear device, not
in demonstration, but as intended—a single bomb capable of destroying a whole city. To my amazement, Bethe also perceived another dimension to the atomic weapon. "It seemed supernatural," he said. "The
bomb made it possible for the Japanese to surrender with honor. They
could still say, 'We fought very well, but there was something which
was far beyond our control—we had to surrender.'"

Bethe asked me if I knew of South African writer Laurens van der
Post's book about his experiences in a Japanese prison camp and his
view of the bomb.[4] When I replied that I did not, Bethe told me that van
der Post had been a colonel in the British army and spent three and a
half years as a prisoner of war. Van der Post wrote that it was the cata-
clysmic nature of the bomb, more like an act of God than of man, that
allowed the Japanese to surrender without dishonor. Bethe's voice
trembled as he told me, "Van der Post is convinced that if the war had
been ended by an invasion of Japan, [the prisoners of war] would all
have been killed."

The ease of surrender was Bethe's final point in support of the decision. Terribly and swiftly, the war was ended, leaving an opportunity
for a solid peace. "When he arrived [in Tokyo Bay] on the [USS] *Missouri*, General MacArthur gave a very moving speech, which some of
the ranking Japanese found very impressive—no revenge, no bad feelings, now we will live in peace." In contrast to the aftermath of World
War I, the resentment that led to a resurgence in militarism was
avoided. Bethe then summarized,

> I have lots of arguments the Japanese were much better off by having
> the two bombs than they would have been otherwise—either by inva-
> sion or by blockade. And this is quite apart from the likelihood that in
> case of an invasion, the Soviets would have occupied part of Japan.

As Bethe concluded his argument, I gazed out at the trees surrounding his home. The snow had been falling all afternoon and the sky was
beginning to darken. Watching the blanket of white gently cover the

tall pines, I experienced a sense of disconnection. We were calmly and safely discussing the instantaneous destruction of two Japanese cities and many thousands of lives. I supposed that when one discusses great wars from any distance, this is always the case. Nonetheless, the idea that the Japanese people were better off with the devastating nuclear attacks was difficult for me to accept. I asked Bethe if saying "the two bombs" meant it was clear to him that the bombing of Nagasaki was necessary. He responded, "Perhaps not. Half a day before Nagasaki, the emperor decided to surrender. But this decision was not known in Washington. Only after Nagasaki was the decision announced over Tokyo radio."

However, Hiroshima, coupled with the Soviet invasion of Manchuria, left the Japanese government in chaos. Did it even have time to react before Nagasaki? The second bomb was scheduled to be used unless Truman halted the mission. The president's command was not needed to drop the bomb but to stop it. Bethe concurred, "You are absolutely right. It may have been a mistake to give the orders in this way, thereby leaving the decision to the commander in the field." I responded that even after Nagasaki, but before the surrender, the horrific firebombing of Japanese cities continued. Bethe reminded me that a third bomb was at Los Alamos, ready to be transported to the Pacific, when the lab received the order not to ship it.

I told Bethe that we could argue about the decision to drop the bomb from various standpoints. The theologian and the general may interpret the same historical facts differently. But the bomb was terrible. In addition to Hiroshima and Nagasaki, the mushroom cloud has become a symbol of our ability to instantaneously kill each other in vast numbers. The image of the bomb encapsulates the horrible reality of modern war and reminds us that we now have the potential to destroy the world. I alluded to the controversy surrounding commemorations of the bomb and suggested to Bethe that people might find it difficult to celebrate the impending fiftieth anniversary. I was expecting him to continue the logic of his argument by pointing to some fact in his notes, or by again emphasizing that the bomb had ultimately saved lives. But instead he said,

I agree completely. The bomb is an evil thing, there is no question. And
my first reaction after Hiroshima was we should never use it again.
Immediately after the surrender I was completely devoted to arms control. But once fission was found and there was a war, it was a foregone
conclusion that the bomb would be made.

This brought to my mind Robert Oppenheimer's famous statement
that the atomic scientists had "known sin." I asked Bethe what his
friend had meant. He explained,

I think it's a correct statement. And he [Oppenheimer] said it in the
context of saying that some day the name "Los Alamos" will be cursed.
It's correct because it brought this evil thing into the world and quite
apart from the Japanese cities, there it is.

Then he added,

I want to mention one more statement which is on the other side.
One of the Los Alamos scientists at the fortieth anniversary said, "It
served them right for Pearl Harbor." And I disagree with that. I think
this scientist had not outgrown the mentality of war and its unreason-
ing brutality. I mention it only to say that not all scientists are angels.

I wanted to ask Bethe more about his colleague and friend Leo Szilard. I told him that Szilard's social and political views held great interest for my young father during his Met Lab days. I found him an intriguing character, in particular, because of his stance against the use of
the bomb after the German defeat. But, I said, from the outset Szilard
must have known the bomb's terrible nature. Bethe replied, "He did.
Well, he was afraid of a German bomb, so he stimulated Einstein to
write the letter, which had very little effect."
    Bethe was alluding to Szilard's early concern about the possibility of
a nuclear weapon. The Hungarian scientist drafted Einstein's 1939 letter
to President Franklin D. Roosevelt explaining that recent scientific
developments pointed to the achievement of a nuclear chain reaction,
which would lead to very powerful bombs. The letter also warned that

the Germans could be working on such weapons. With Edward Teller as his "chauffeur," Szilard had traveled from New York City to eastern Long Island, so that Einstein could review and sign the letter. Linus Pauling reported that, late in his life, Einstein confided in him, "I made one great mistake in my life—when I signed the letter to President Roosevelt recommending that atom bombs be made, but there was some justification—the danger that the Germans would make them."[5]

Bethe explained to me,

> Szilard always looked far ahead in his thinking, and he said, "Well, this is terrible, nuclear weapons can be made, the Germans surely will try to do so, we have to do it first." And then, when it came to the crunch, so to speak, he said, "Well, the Germans haven't had it before they surrendered, we have it, let's not use it." He didn't think of all these points, he didn't see that far in the future.

Did Bethe mean that Szilard's moral opposition to the bomb's use was based on an incomplete understanding of the possible consequences had it not been used?

> Yes. Well, of course, he thought that a demonstration would be enough. I believe he did not realize the fanaticism of the Japanese military at that time. He certainly couldn't know that it was all up to the emperor. He thought that convincing enough Japanese people would be enough, and maybe it would have been, I can't tell. I don't know how the demonstration would have been done. I think he did not realize the supreme importance of the emperor. Well, maybe we would have invited the emperor to witness it, but the Japanese might have found ways to keep him home. Things were sufficiently uncertain, sufficiently confused, sufficiently urgent, that you couldn't foresee all these points, not even Szilard.

If, as Bethe said, the bomb was evil and war is evil, how did he as a scientist, who brought this evil into being, view his own accountability? It seemed to me that implicit in his statement was a value system of some kind, an absolute judgment of what is good and what is evil. Where did he stand in reference to this? Bethe answered, "Yes, well, I take the pragmatic point of view. What would do the least harm to the Japanese?"

I remarked that he seemed to be making the lesser of two evils argument, and his response moved the conversation from his assessment of Hiroshima to the postwar nuclear situation.

> That's right. Well, my reaction after Hiroshima was, this shouldn't be repeated. But that is really not a good way to say it, because if there were nuclear war between nuclear powers like Russia and the United States, it would be an entirely different matter. It would not be the end of the war, it would be the beginning. And using only one percent of the arsenal, it would destroy both countries, and presumably also Western Europe and, in Soviet times, Eastern Europe. And there were fortunately some nonaligned countries, and if we leave them in peace, in such a catastrophe, maybe civilization could be built up again from South America and South and East Asia.
>
> Anyway it would be a totally different matter from Hiroshima and Nagasaki. And from the word go, that is, immediately after the Japanese surrender, I was completely devoted to arms control, to preventing nuclear war. Well, it's not so easy to prevent nuclear war, and I think on the other hand that statesmen on both sides and also [the] neutrals have realized that a war between nuclear powers is just impossible. The consequences are just so staggering, and I think any rational statesman will subscribe to that. Gradually, in the course of fifty years, the responsible statesmen have realized that nuclear war is impossible. And once you have a war between the great powers, it is very likely that it would become nuclear.
>
> So, and here again I say it is *not* a question of morality, but it is a question of understanding. Just as I think my reasoning about dropping the bomb is a question of understanding, long after the fact, what the story really was. And fortunately, statesmen have understood it. [That the bomb is] a peril and a hope was, of course, very much the idea of Niels Bohr, who probably thought about these matters more than anybody else before the end of the war. He tried to make Roosevelt and Churchill understand, but you know that his conversation with Churchill was a total catastrophe.

Bethe was referring to Danish physicist Niels Bohr's unsuccessful wartime attempts to persuade Roosevelt and Churchill that they should negotiate an agreement with Stalin for postwar international control of atomic energy. Bohr saw a historic opportunity to lay the groundwork for a new open world, before the bomb was unleashed. He was convinced that implicit in the existence of this terrible new destructive force was the seed for a radical rethinking of humankind's affairs. Bohr summarized the purpose of his meetings with Roosevelt and Churchill: "The grave responsibility, resting upon our generation, [is] that the great pioneer effort, based on the recent advance of physical science, is used to the benefit of all humanity and does not become a menace to civilization."[6]

Bethe, by using the phrase "a peril and a hope," was attempting to convey to me the deep paradox, or "complementarity," that Bohr perceived in nuclear weapons. And Bethe was asserting that it took statesmen a half century to understand the basic principles of arms control, which Bohr had labored to communicate to Roosevelt and Churchill during the war. He explained, "When Niels Bohr came to Los Alamos late in 1943, he asked Oppenheimer, 'Is it big enough?' Namely, big enough to make war impossible?" And what had Oppenheimer told Bohr? "I think, 'Yes.' And then he and I and Feynman gave him details."

During the war, Bethe never imagined the tremendous buildup of nuclear armaments. He once reflected, "When the Los Alamos laboratory had built the first nuclear weapons, we scientists thought that if national arsenals ever included them at all there would be very few, perhaps a few dozen."[7] Now he told me,

> The numbers of weapons which were accumulated by both sides were absolutely absurd. Twenty thousand, thirty thousand—what do you do after the other country is already destroyed? And it makes absolutely no sense to have such numbers, explained only by the *mindless* arms race. So the arms control movement was a big movement, and after a while, I think the President's Science Advisory Committee, on which I served under Eisenhower, convinced Eisenhower that arms control was necessary and Kennedy didn't even need to be convinced. And so, through the years, there were lots of discussions inside the arms control

community. And the first thing we wanted was control, that is, put limits, and one of the things we said at that time is that an antiballistic missile is a *bad* idea, because the offense can always build more.

Now, I'm for arms control, I'm for arms reduction, but I'm against complete elimination of nuclear weapons. The reason being that there are countries which have *not* understood that nuclear war is impossible. And two of them are Iraq and North Korea. And Iraq's aim was to wipe out Israel, apart from dominating the Arab world. North Korea's aim was to wipe out South Korea. Such countries exist, and such countries are likely to continue to exist, and therefore it seems to me that we need to keep some nuclear weapons. Maybe fifty, maybe fifty for Russia, maybe ten or twenty for China, France, and Britain. [But] we are still talking about *thousands*. There is no earthly reason for the thousands.

And so deterrence is necessary. And very important, of course, is nonproliferation. So we'll have trouble with the extension of the non-proliferation [treaty]. We had trouble before with India and Pakistan. We had trouble before with Brazil and Argentina, they have given in, they have stopped their development. And one of the very unhappy features is that having nuclear weapons is supposed to be a matter of prestige. And that concerns particularly India and Pakistan.

As our afternoon's conversation drew to a close, Bethe gave me the papers containing his notes along with a volume of his collected works, and we arranged for a follow-up interview the next day. The Bethes invited me to join them and another Cornell professor and his wife for dinner. Later that night, back at the hotel, I slept badly. There was a noisy party down the hall and my mind was racing. If, for Bethe, the bomb was an evil, did that mean he believed in *the good?* His logic would seem to require it. What were the values by which we could judge such questions? Finally, early the next morning, I fell into a deep sleep. I awoke a few hours later to prepare for the follow-up interview. While showering, I thought of my parents and of unanswered questions. And I realized that after speaking with Bethe, I was experiencing a sense of relief. It did not matter that I sometimes disagreed with him. He listened carefully to my questions and answered them patiently and thoughtfully—it was the dialogue that counted.

I had some breakfast, and soon Hans Bethe drove over to the hotel for me. On the way back to the house, we passed a construction site, and he told me this was where he and Rose would soon move—a retirement community called Kendal—"So they can take care of us," he said. Pulling into the garage, he confided that he was afraid of moving after living in their home for fifty years.

Rose Bethe joined us, and we discussed how I might endeavor to have an article about our conversations published, as I had no experience in these matters. During the chat, I said it could be a worthwhile article—Hans Bethe did not easily fit into the comfortable categories usually associated with the use of the bomb, arms control, and the problem of nuclear weapons. He told me with a chuckle, "My wife said of me that I'm a dove, but I'm a tough dove."

When we continued with the actual interview, I told Bethe that a question had arisen from our previous day's conversation—one that was more a question to myself than to him: Why was it important for me to speak with him? I attempted to answer it by thinking out loud. I said that when my mother died, she had asked me to take care of my chronically ill father. And, at the end of his life, my father seemed to be telling me that although he had never returned to weapons work, he remained deeply troubled by the bomb. By showing me how bad he felt, he had left the question of the bomb as a legacy. I told Bethe that the whole situation was strange and sad—it was as if my parents were passing on to me something they wanted me to do. And, I said, there was some deep logic in my having these conversations with him.

I did not want to, but I began to cry. Fearing that the great scientist would find my ramblings incoherent, I asked him if I was making any sense to him. He responded, "Sure." Through my tears I told him that if someone were to ask why I had traveled all the way to Ithaca to see him, I wouldn't have any rational explanation, beyond that I wanted to talk to him about the bomb. He looked at me sympathetically, perhaps a little quizzically, as I thanked him for helping me to move the intergenerational question along. I told him it was a difficult thing for me to do. "Yes, indeed," he said, "I realize that." Then I saw him reaching toward me with something in his hand and heard him say, "Have some chocolate."

I ate the chocolate. Then I returned to my list of follow-up questions.

I told Bethe that I could accept that for him the decision to use the bomb was the least evil among several real and savage choices. However, I remained deeply troubled by the atomic bombing of defenseless civilians. Even though the targets had ostensibly been chosen because of their military significance, this was a thin veil to cover what everyone must have known would be a devastating attack on noncombatants. Bethe answered,

> The Nazis did it in the Blitz against Britain, and the Allies did it with increased intensity. Then it went to the firebombing of German cities, and especially of Japanese cities, like Tokyo. In the course of the war you didn't think about the victims anymore—so everything was justified in war. It made the warriors brutal, and the warriors included such thoughtful people as [General George C.] Marshall, Stimson, Roosevelt, and scientists. Then the other side, other remark I would make to this, is that, really, the atomic bomb has made war of the old type impossible between nuclear powers.

I asked the question I had been thinking about the night before. Bethe had told me that he thought the bomb was a necessary evil. I wanted to know if he believed in an absolute good. He was silent for a few moments and then responded,

> Take the atomic bomb. It's an evil thing, as I said. When it is used, if it were used again, it would be a terrible evil. And yet it was good in the sense of saving people's lives at the end of the Second World War. It saved American lives and it saved Japanese lives. Is it good? It certainly wouldn't be in an all-out nuclear war. But at that particular moment, it probably was good. It was so considered by all the servicemen in the Pacific. And so, and in retrospect, as I tried to say at length yesterday, it was good for the Japanese. So I guess in most cases, there is no absolute good.

Was he saying that this was a relative good?

> Right. Socialism is probably good for the worker, it's probably necessary, and yet in the form practiced in the Soviet Union, it was certainly bad.

So I think from my life's experience, I would say that good is relative, most of the time. And yet, we all feel, Do not do unto others what you don't want to be done against you. And that's absolute good. So I think this question will remain unsolved just as it remained unsolved for Socrates. He said if somebody knows what is good, he will do good. But very often, we don't know.

I asked if he was now optimistic about the world's future. He replied, "Optimistic in the sense that I believe there will not be a major nuclear war. There may be occasionally a single bomb, I don't know." Did he think that human beings would be able to keep pace with their technological inventions?

Yes, I think so. I guess I am an unreconstructed member of the Enlightenment. One of the leading philosophers of the Enlightenment—Condorcet—believed that man can be perfected indefinitely. And he wrote that down while he was imprisoned by the [French] revolutionaries and he was finally put to death. It takes a lot to have that much optimism under those conditions.

Did Bethe conclude that reason would ultimately prevail? He replied that he did and that historically it had done so in the Western world. I remarked that there was certainly a dark side to that—a great deal of destruction had been caused by the progress of the West. He responded, "Absolutely, but you see, the Allies prevailed in the Second World War. And Communism died of its own weight in Russia." Was he saying that it was through reason that our current dilemmas would be solved? Bethe concluded, "Yes, that's my belief. And I may be wrong. I mean, there are always setbacks."

We brought the interview to a close. Bethe had been generous with his time, and I had to catch an afternoon flight to Maine. As I packed my recording equipment and thanked Bethe, he replied that he had enjoyed talking to me. "I never bothered to think about absolute good and relative good," he said. I told him that although I did not claim that the idea was original with me, I had come to the distinction between absolute good and relative good while speaking with him. I added that I sometimes think by talking. He replied that Niels Bohr, the century's second greatest physicist, could only think by talking. When I asked who had been the century's greatest physicist, Bethe responded, "Einstein. But in many ways Bohr was greater because he was very sensible about life in general and politics. And he had a tremendous school, which Einstein didn't have."

Bethe's final remarks only hinted at Niels Bohr's important role in the atomic drama. As Bethe and many of his colleagues would subsequently tell me, the great Danish theoretician not only taught a generation of physicists, his beliefs about the larger social and political meaning of nuclear weapons had a profound impact on the thoughts and actions of the atomic scientists, in particular, Robert Oppenheimer.

After meeting with Hans Bethe in Ithaca, I wrote an article about my conversations with him, and my memories of my father, which was published six weeks before the bomb's fiftieth anniversary.[8] Rather than feel my work was complete, I sensed it was just beginning. When I told Bethe that I was thinking of writing a book, he encouraged me to move forward and suggested several of his colleagues, both atomic scientists and others, with whom to speak. Thus I widened the circle of dialogue on the questions that were troubling me and crossed the boundary from personal to public.

On the actual anniversary days of the atomic bombings of Japan, August 6 through August 9, 1995, I attended a peace retreat in Santa Barbara, opening the conversation in yet another way.

✳ A Thousand Cranes

August 6, 1995
50th Anniversary of Hiroshima
Santa Barbara, California

At a peace retreat cosponsored by La Casa de María and the Nuclear Age
Peace Foundation. In the afternoon, we dedicate the peace benches and grotto
beneath the "tree of faith" of the Immaculate Heart Center. Here, in Santa
Barbara, there now exists a tiny garden commemorating the victims of
Hiroshima and Nagasaki. Under the protection of the ancient eucalyptus
branches we gather to pray and ponder the meaning of those world-changing
events of fifty years ago. Physicist Walter Kohn speaks at the dedication. I am
sorry that I do not introduce myself as he will not return for the rest of the
retreat. He talks about the violence in all of us.

This evening, in the chapel, Stella Matsuda performs "Dance of a Thou-
sand Cranes . . . Up from Ashes," which she created in honor of the bomb's
victims. In front of the high, vaulted window hangs a life-sized wooden cruci-
fix. The altar is strung with one thousand origami cranes. Below them, Stella
dances with elegant, birdlike gestures. At the end, she stands suspended mid-
movement, body trembling.

The crane's broken wings perfectly reflect the outstretched arms of the
crucified Christ.

The tree of faith was so named by Noreen Naughton, IHM, then
director of the Immaculate Heart Center. Artist Irma Cavat and landscape architect Isabelle Greene designed the peace benches and garden,
christened Sadako Peace Garden in memory of a twelve-year-old child
from Hiroshima who died of radiation-related illness ten years after the
atomic bombing. Sadako Sasaki folded paper cranes as a prayer for
recovery and for world peace. "I will write peace on your wings," she
said, "and you will fly all over the world." She died before making the

thousand cranes, and her classmates completed them as a symbol of hope.

Stella Shizuka Matsuda is a third-generation Japanese American whose parents, Michio and Taneko Nakadate, were born in Hawaii. In 1942 Stella lived with her parents and brother in the Los Angeles area, where her father was a dentist. Shortly after Pearl Harbor, Secretary of War Stimson decided to evacuate persons of Japanese ancestry from military areas. Stella was five years old when President Roosevelt signed Executive Order 9066, authorizing this action. Within two weeks, California, Oregon, and Washington were declared strategic regions. Over 110,000 Japanese Americans, more than two-thirds of whom were American citizens, were forced from their homes and businesses and imprisoned in ten camps.

Stella's family had to move in a matter of weeks, so her father left his dental equipment in safekeeping with neighbors. The family was first sent to Santa Anita Racetrack, where they lived in the horse stables. Then they were relocated to the Poston, Arizona, internment camp, where Stella's father established a dental clinic. On returning home in 1944, Dr. Nakadate discovered that all of his equipment was gone.

CHAPTER 2

Edward Teller,
High Priest of Physics

On December 1, 1995, I traveled to Stanford University to meet with Edward Teller at the Hoover Institution on War, Revolution and Peace, where he is a senior research fellow. Left to my own devices, I would not have attempted to interview Teller. I am from a liberal background, and his name held many negative associations for me, from the matter of J. Robert Oppenheimer to the development of the H-bomb, the nuclear freeze, and the Strategic Defense Initiative. My concern was that my biases would make an interview too difficult. However, Teller had been one of the first on Hans Bethe's list of those with whom I should speak. I wrote to Teller describing my project and saying that Bethe suggested I interview him. He sent a cordial reply, granting me the interview. Then, during the process of setting up the appointment, I had several pleasant telephone conversations with his assistant, Patricia French. Nevertheless, when I walked through the front door of the Hoover Institution, I was uneasy.

As I arrived at his office, Teller was ending a meeting with a young European man who requested that Patricia French photograph them together. She then led me from her outer office into Teller's and introduced us. She offered us some lemonade and left the room to get it. I had barely taken my seat when the eighty-seven-year-old physicist said emphatically that he did not know exactly why I had come but there were some things he intended to make clear to me from the outset. First, he said that the United States' expenditure on nuclear weapons was a smaller percentage of the total defense budget than commonly believed, making the point that criticism of spending on nuclear weapons was overblown. I sat very still, unsure of whether I should move or speak. I had not asked his permission to tape the interview or even taken out pen and paper. As he recited the defense dollars, I indicated that I wanted to take some notes and needed my writing material. He replied that I did not need to take notes.

Immediately on meeting Edward Teller, I had two reactions. The first was that I felt intimidated and dared not interrupt him. The second was that he seemed intent on persuading me of the rightness of his views. He spoke emphatically, his Hungarian-accented voice rising to punctuate crucial ideas and falling dramatically to almost a whisper at the end of key arguments. To add force to his words, he wielded his heavy walking stick. As I oriented to the room, I noticed laces protruding from the top of one of his black cowboy boots and assumed that they were part of a prosthesis. I knew that in the late 1920s, while a student in Budapest, he had lost a foot in a streetcar accident. Behind Teller's head, I saw a bust of Abraham Lincoln.

After several minutes, when there was a break in his monologue, I obtained his permission to record the interview. After making his point about defense spending, Teller turned to an issue that was of more immediate interest to me—the atomic bombing of Hiroshima. On that subject Teller told me he had regrets, "a weak regret" and "a strong regret." The weak regret was related to a July 1945 letter and petition that his friend and countryman Leo Szilard had sent from Chicago's Metallurgical Lab to him in Los Alamos. The strong regret had to do with a request that Enrico Fermi had made of him a few weeks earlier.

Before receiving Szilard's letter, Teller had been in Chicago. He was therefore aware of the Franck Report and other discussions taking place at the Metallurgical Lab regarding the political, social, and moral implications of the atomic weapons. Szilard's correspondence to Teller contained the petition that he hoped to have circulated at all the Manhattan Project laboratories and then forwarded to President Truman.

I had read Teller's comment that he had been "inclined to sign the Chicago petition, but could not circulate it without checking the matter

Figure 10. Edward Teller,
Los Alamos, circa 1944.

with Oppenheimer." He was now telling me he had felt
obliged to inform the Los Alamos director of Szilard's communiqué
before doing anything. He said that Oppenheimer's response was that
the petition should not be circulated at the laboratory because the leaders in Washington knew what they were doing. Teller added that
Oppenheimer told him it was not the scientists' job to try to influence
the politicians and the military on matters about which they had little
understanding. Teller has also recorded his recollection of the incident:

> Oppenheimer immediately offered several uncomplimentary com-
> ments about the attitudes of the involved Chicago scientists in gen-
> eral and of Szilard in particular. . . . My predominant feeling fol-
> lowing our conversation was relief—I did not have to take any
> action on a matter as difficult as deciding how the bomb should
> be employed.
>
> Later I learned that shortly before that interview Oppenheimer
> not only had used his scientific stature to give political advice in
> favor of immediate bombing but also had put his point of view

forward so effectively that he gained the reluctant concurrence of
his colleagues. Yet, he denied Szilard, a scientist of lesser influence,
all justification for expressing his opinion.[1]

Teller was alluding to Oppenheimer's membership, along with his
colleagues the Nobel laureates Enrico Fermi, Ernest O. Lawrence, and
A. H. Compton, on the Scientific Panel advising the Interim Committee
of the War Department. In May 1945 Secretary of War Stimson had
formed the committee at the urging of advisers who believed it was essential that immediate attention be paid to the postwar implications of
nuclear energy and nuclear weapons. Stimson asserted that the bomb
project "should not be considered simply in terms of military weapons,
but as a new relationship of man to the universe."[2]

Teller wrote to Szilard, "Since our discussion I have spent some time
thinking about your objections to an immediate military use of the
weapon we may produce. I decided to do nothing."[3] Teller told me that
although he blamed Oppenheimer for not allowing the petition to be
circulated, he nonetheless regretted having been persuaded by him.
However, he said it was a "weak" regret because he had not been offi-
cially asked for input about the use of the bomb.

Teller contrasted these misgivings with his strong regret about the
conversation he had with Enrico Fermi several weeks before receiving
Szilard's letter. That discussion concerned a possible demonstration of
the bomb. Teller said that his Italian-born friend had consulted him
about how the bomb might be technically demonstrated and that he
regretted having given insufficient thought to this problem. In this case
he had been asked to look at a particular technical question and had not
done so. He explained that he should have come up with a concrete plan
that the scientists could have presented to the leaders in Washington.
This, to Teller's mind, could have been a demonstration of an atomic
bomb over Tokyo Bay, where the emperor and the Japanese people
would have seen it but the danger would have been minimal. He envisioned parachuting the bomb from an altitude of six miles, having
primed it to explode when the plane was away. At a sufficient height, he
said, the only casualties would have been people blinded by looking

directly at the blast.

I understood Teller to say that Fermi, in his capacity as a member of the Interim Committee's Scientific Panel, had specifically asked him to explore possible alternatives to the military use of the bomb and that he had not given the problem adequate attention. On May 31, 1945, after discussing possible targets and effects, Stimson expressed the committee's conclusion, on which there was general agreement:

> That we could not give the Japanese any warning; that we could not concentrate on a civilian area; but that we should seek to make a profound psychological impression on as many of the inhabitants as possible. . . . The secretary agreed that the most desirable target would be a vital war plant employing a large number of workers and closely surrounded by workers' houses.[4]

The Atomic Energy Commission's official history concluded that during lunch on May 31, perhaps ten minutes were spent generally discussing an issue Lawrence had raised at the Interim Committee's morning session: "give the Japanese some striking but harmless demonstration of the bomb's power before using it in a manner that would cause great loss of life." However, the historians reported,

> Oppenheimer could think of no demonstration sufficiently spectacular to convince the Japanese that further resistance was futile. Other objections came to mind. The bomb might be a dud. The Japanese might shoot down the delivery plane or bring American prisoners into the test area. If the demonstration failed to bring surrender, the chance of administering the maximum surprise shock would be lost. Besides, would the bomb cause any greater loss of life than the fire raids that had burned out Tokyo?[5]

On June 16, 1945, the four scientists on the panel reported, "We can propose no technical demonstration likely to bring an end to the war; we see no acceptable alternative to direct military use."[6]

I wondered if Teller's demonstration idea was similar to the Franck committee's recommendation to use the atomic bomb on an uninhabited island. He told me that one objection to the Franck Report had been that if something went wrong, if the bomb did not work, such a demonstration would make matters worse. When I asked him how his plan differed, he explained, "Because it would be unannounced. We drop a bomb, all right! It does not go off, then we have done nothing! Here the failure is: it didn't work, do it again or forget it. But it did not do any damage."

When I questioned how, without an understanding of nuclear physics, the emperor and the Japanese people could have comprehended the significance of such a demonstration, he continued,

> Look, I have seen it. Here you are at 6:00 A.M., it's dark, and there is light for five minutes! They [would] know something very strong, something very effective has happened. And ten million of them have seen it, so it was over a big area. It would not be proven, but it would be plausible that the same thing, used for destructive purposes, could destroy.

I asked Dr. Teller what, in the best of worlds, he envisioned as the possible consequences of such a demonstration. He replied,

> Look, let me tell you, in the most real of worlds, and taking into account what you have read and I have read: ten million Japanese would have seen it, [Emperor] Hirohito would have seen it. Hirohito *heard* of Hiroshima. Hirohito instead would have *seen* the effect over Tokyo Bay. We would have explained, "This is what happened," and next time we would use it over a city. I think the inducement of Hirohito to do something would be comparably strong, if not stronger [than it eventually was]. And furthermore, he would have an easier job, because he would talk to Japanese who have *likewise* seen it.*
>
> Incidentally, if that would have happened, it would have had a different, but a comparable, effect on the Soviets. Well, not as strong, because nobody would have been killed. The very fact that it ends the

---

* Harold Agnew, a Manhattan Project physicist who flew the Hiroshima mission, told me in a January 1999 telephone conversation that Teller's idea was nonsense. Six miles was thirty thousand feet, as high as they could fly, and it would have been difficult, even impossible, for the plane to get away in time. In a February 1999 letter, Agnew added, "We also only had 2 bombs. The ability to deliver bombs 3 days apart I believe gave the impression we had lots. What would we have done if they told us to jump in the lake after the demonstration? Having only one bomb to use would not in my opinion [have] convinced [the Japanese] to quit."

war by a demonstration would have been a shock, and furthermore, something that could not have been just covered up. And you know the moral force of saying, "We have ended the war without killing a single person." That would have been so strong a statement that it would have influenced many people. Not Stalin, but many people, many people including many Communists.

In what way, I asked Teller, would the Communists have been influenced? "That their opponents, the United States, were not all that horrible." His answer surprised me because it implied that the use of the bomb had run counter to America's interests regarding postwar relations with the Soviets. Yet some scholars argue that the United States dropped bombs on Japan precisely to subdue the Soviets. I asked Teller if he was saying that the atomic bombings actually had a negative impact on relations between the United States and the Soviet Union. He quickly responded, "Look, look, it had an effect, it had an effect toward winning the cold war. But it had a warlike effect, which being a warlike effect, had also necessarily contained a very strong negative emotional component."

It so happened that Teller was then reading Gar Alperovitz's recently published work on the decision to use the atomic bomb.[7] He agreed with Alperovitz on one point but was otherwise critical of the book.

Probably the war could have been ended by [the United States] being more open on retaining the emperor sooner. And you know it so happens that my suggestion would have been a fortunate one. The emperor would have seen it, and furthermore he would have talked to the Japanese, many of whom also would have seen it. And so he would have been in an excellent position. But [Alperovitz's argument] that our emphasis on influencing the Soviets was somehow very wrong, with that I don't agree. That was an opening of the cold war, indeed, it was indeed a beginning of a lot of enterprise which amounted to stopping the imperialistic trend in the Soviet Union—the trend to force on other countries their form of government. And to my mind, that was a necessary mission.

But let me tell you, in a way, the *strongest* argument I have, all right? Not Russia, not Communism, but today, there is in my opinion, a very dangerous antiscientific movement afoot. I [will] illustrate this to you. I give a lecture on modern biochemistry—a public lecture. Without exception, I get one question from the audience: "How can this be misused?"

I came to this country in 1935, sixty years ago. At that time nobody would have asked that question. That nobody would have asked that question, that there was a confidence in people, that on the whole scientific developments were desirable, is the reason of the strength of this country today. The negative feeling about science—that you ought to be afraid of it—is a strong reason for a potential weakness of the United States. Had we avoided that, had we said, "Science stopped the Second World War, without the loss of a single life," it would have been good for the proper support of science.

The discussion had quickly moved from Teller's fifty-year-old regrets about his friends Szilard and Fermi to his arguments against what he considers today's dangerous antiscience stance. Although he made connections between the atomic bombings and this attitude, he nonetheless was highly critical of those who find in the use of the bomb justification for condemning science and scientists. I was not sure of the logic of his argument, but his remarks raised a question in my mind. Because he seemed to be making a direct connection between current negative feelings about science and the atomic bombings, I asked him if he had ever wondered what might have occurred if the discovery of fission had not coincided historically with the rise of Nazism and Communism.

Teller closed his eyes and put his hand to his temple. The short pause I was expecting before his reply grew into a long silence. Although I had begun to relax, I still felt considerable anxiety and laughed inwardly at my predicament. I thought, either he is controlling his anger because this is the stupidest question he has ever heard, or he is thinking about what I asked. Hoping he would not notice, I held my breath as he opened his blue eyes, peered out from under his thick brows, and fixed his gaze on me.

Look, I cannot tell you what would have been. I can tell you what I actually knew and how I reacted. I knew of that toward the end of January of 1939. We had a conference in Washington, to which Niels Bohr came on the invitation of [Russian-born physicist George] Gamow. And on the evening before the conference opened, Gamow, who was close to Niels Bohr, called me, and told me on the phone, "Bohr has gone crazy, he tells me that uranium splits." Next morning I already knew. I had heard about Fermi's experiments, and I figured out what he was talking about. He talked about it next morning. There was general agreement that this could be very dangerous and that we'd better not talk about it. You know, this story [of the discovery of fission] came to me in the middle of terrible worries of how the Nazis ever could be stopped! And I must confess, I never thought of the question of how it would have been, had it been otherwise.

As Teller continued, he again brought the discussion to the present-day fear and mistrust of science, in this case, the application of science in the production of small nuclear warheads.

I'll now tell you, I may be wrong, I think that had the Nazis not been around, we would have developed it. We would have developed it openly, and the question would have arisen, under less hysterical conditions, how misuse can be prevented. Whether we would have succeeded with it or not, I do not know. But I want to draw your attention to a simple fact, that today we have fortunately a repetition. Soviet Union no longer exists in its imperialistic form. We have now the possibility to open up and to find the way to regulate this thing. And I must say that the whole discussion has been colored by the hysteria and that people, instead of talking about the most probable, are tending to talk about the worst.

For instance, I claim that to know about it would be good, to develop small nuclear explosives to make the present methods of mass destruction, like tanks, obsolete, because you would have small, cheap, easily delivered things that would destroy a tank from a distance of one hundred feet. Immediately people say, "We can't do that because of terrorists." The point that such a weapon might be easily smuggled is not wrong, but people immediately, automatically, concentrate on what is easiest, not on what is most difficult.

So I claim the slow development where openness would be the route and where we would try to find out and could find out who is doing funny things, could probably lead to a very stable situation. And that could have happened fifty or sixty years ago. And what is more important, it could happen now. But even now it's not happening. Because people tend to think about the worst case.

Teller's reference to small nuclear weapons indicated that he believes the institution of war, as we know it, will be with us for some time to come. I recalled the argument that nuclear weapons mean humankind must end war or face the possibility of complete annihilation. I asked Teller if he meant that war is a given. And, if that were the case, was he saying that nuclear weapons development must continue? He responded,

> I am not answering this. That war is a terrible thing and has been even in times of Genghis Khan, it's quite clear. I am not saying that war is going to be over soon, and I'm certainly not saying that war can be stopped by disarmament agreements. I am saying that wars can be stopped only by a careful, gradual development, and I am also saying, however, the need for such a development, and, in principle, the possibility of such a development, is now much bigger than ever before.

Teller was presenting me with an argument for continued weapons development after the cold war. I was trying to understand more clearly why, if his advocacy during the cold war had been because of the Soviet threat, there was now no apparent change. I told him that my reading of his earlier works was that he expected there would be a real nuclear confrontation between East and West.

> I didn't say "would," I say "could." I differed from most people, *not* in predicting that such a thing might happen, but in restricting the ways in which it might happen. It did *not* come to pass, in my opinion, because there was no time where the Soviets could have attacked us without the, not possibility, but probability, of they themselves going under. Probably we too, but they too.

Many times during the interview Teller asserted that people oppose such technological development because they focus on the "worst case." I understood this to be directly related to his point about the irrational fear of science. But did he not see some justification for this fear? After all, science and technology are often used in ways that people judge to be counter to moving us forward, realizing our potential, or doing us good. He answered, "Look, let me take two very different things, both of which have some relation to an answer. One is that throughout all written history the killing of people was never limited by the ability to kill people but always by the amount of intention to kill people." He asked me if I knew what happened under Genghis Khan. When I answered yes, Teller said that he knew better than I.

> Because after his death, the Mongols under two generals descended on Hungary and killed 90 percent of the Hungarians in a few weeks. You know what they did was—they would attack a city, kill everybody, then leave, then come back in a week, and kill those who crawled out from the weeds. That happened to Persia, it also happened a little later to Hungary. The atomic bombs killed 150,000 people out of the fifty million that had been killed in the Second World War. The limitation is not in the ability but in the intention. That is one thing I wanted to call your attention to. So there was reason to fear at *any* time.
>
> The other point I want to draw your attention to is that science in the last century was understood and liked by the intellectuals. Then, at the beginning of the twentieth century, two scientific developments occurred that were beyond 99 percent of the intellectuals: one was relativity and the other one was quantum mechanics. That looked like mathematical tightrope dancing.

Was he saying that the new theories were beyond the intellectuals' understanding?

> [Beyond their] *capabilities* to understand. They *knew* it was wrong, and somehow they were informed it was not. They did not like it. They did not understand it and they did not like it. Now that happened in two chunks, or in three. One was Einstein in 1905, then came general relativity in 1915, then came quantum mechanics in 1925. All of this was far beyond popular understanding and far beyond the understanding of the intellectual leaders. What you don't understand, you don't like. Then,

out of that not understood and disliked science, came Hiroshima. What
came afterwards was a confluence of lack of understanding and horror.
The two together produced absurd policies.

When I replied that his argument was interesting and that I could
understand his logic but had a further point, Teller quickly interjected,
"Look, let me tell you what state you are now in. You said you see my
logic, right? I think you do, but you don't *feel* my logic." It seemed that
Teller was applying to me the same analysis that he applied to the intellectuals he was discussing. I did not feel his logic, did not truly comprehend it in the same way the intellectuals could not feel relativity or
quantum mechanics. Teller continued, saying that science had two enemies:

> One was the Church, the other is the modern intellectual. The intellec-
> tuals I am *not* talking about are the people who make an honest intellec-
> tual effort but who, many of them, overestimate themselves. And
> where they have not understood something, they have a feeling, here
> something is wrong, not in me, but in what I haven't understood.

I then asked Teller if we were not talking about two separate realms,
the scientific and the social, to which he responded, "Entirely true." I
continued by arguing that a physical scientist, an intellectual in the
physical scientific realm, looks at things differently and draws different
conclusions from someone examining the social, political, or spiritual
meanings of events. Therefore, the interpretation of a social scientist
will differ from the kinds of conclusions drawn by a physical scientist.
He responded,

> Of course, of course. Except that the social attitude should be about all
> these things, and there science, or then science, should have the same
> value and meaning as music—as music, provided that they [the intellec-
> tuals] are not musical, but they understand that it has a power that is
> wonderful and important, even if they [have] not understood. Remem-
> ber I am saying that a shock of scientific, technical roots—the atomic
> bomb—has been put already at an original disadvantage by its having

come from an area that had already excited some tremors, not having understood it.

He was bringing up a point that I struggle to comprehend—the relationship between science and its products. I can appreciate that for the physicist, deep mental interaction with the subtle workings of nature can be a kind of spirituality, a beautiful and awe-inspiring endeavor. But once I move into the realm of the application of scientific knowledge, especially the making of weapons, the analogy between science and art breaks down. Although Teller claimed that knowledge and its application were separate, he seemed to be attaching equal value to them. Surely pure physics and its applications are linked, but what links them is choice. And such choices are not made on the level of physical science but in the political, moral, and social realms.

I was very interested in what he had written about the scientists who opposed the development of thermonuclear weapons: "Can lack of knowledge ever contribute to stability or safety? The hydrogen bomb controversy marked the first time that a large group of scientists argued for remaining ignorant of technical possibilities."[8] Did he think that this was the first example of scientists being unwilling to pursue certain kinds of knowledge? He responded that although he was not sure it was the first case, it was a striking case. I replied that it had been valid to ask why the H-bomb should be built. After all, the fission bomb was destructive enough. Why would we want a bigger weapon? Teller replied,

> Look, let me put it in perhaps a not permissible way, in words of one syllable. I do not want the hydrogen bomb because it would kill more people. I wanted the hydrogen bomb because it was *new*. Because it was something that we did not know, and could know. I am afraid of ignorance. As it turned out, when the hydrogen bomb was planned we began to debate about it, everybody, including Fermi, emphasized what you are now saying—"It's bigger, why should it be bigger?" That was the debate that occurred in '49.

Teller was referring to a 1949 statement by Enrico Fermi and I. I. Rabi: "The fact that no limits exist to the destructiveness of this weapon makes its very existence and the knowledge of its construction a danger to humanity as a whole. It is necessarily an evil thing considered in any light."[9] The two physicists were members of the General Advisory Committee (GAC) of the Atomic Energy Commission, headed by Robert Oppenheimer. In 1949, after the Soviets detonated an atomic device, the short-lived American monopoly on nuclear weapons ended. This set off a contentious, top-secret debate, within government and among scientists, regarding the hydrogen bomb. The GAC's October 30, 1949, report argued against the development of the hydrogen bomb but recommended the accelerated development of atomic weapons.

Teller, recalling those early years of nuclear weapons research, told me,

> People were overly impressed by the argument of size. Let me tell you, the bombs on Hiroshima and Nagasaki were approximately seventy kilotons; the bombs today, most of them stockpiled, are bigger. I'm not sure I'm allowed to tell you how much bigger, but the fact is, not terribly much bigger. And the reason for that is *not* that it's forbidden. The reason for it is that the military have recognized what is the most useful size, considering delivery, et cetera. The hydrogen bomb is bigger, but the most important point about it is that it is different. And its difference can be exploited.*

I asked Teller if he meant that it was not just the knowledge that was new but also the way it could be used. He replied, "In this case, we were not talking about new knowledge at all in the scientific sense. We were talking about new knowledge in the technological sense—how to, by what trick to turn that knowledge into something usable." I then wondered if he meant that, regardless of the application, such a develop-

---

* In his November 17, 1997, correspondence with me, Herbert F. York recalled, "Both Lawrence and Teller were indignant over the GAC's allowing their 'political and moral views' to influence, even overshadow, their technical judgments, and that they had in fact adjusted the technical report to reflect their 'feelings' in the matter."

ment is "almost good," simply because we can figure it out. He immediately answered,

> Good, not almost. Good. Look, the scientists, by giving you the tools, are not responsible for the use of these tools. But they are responsible for the effectiveness of the tools and for the understanding of the tools. I and you as citizens are responsible for selecting the decision makers who will then use whatever can be used in the right way. And these functions should be separated. My position is that knowledge is good and must be separated from the application of knowledge. And anything that can be applied can be used or misused.
>
> Our government does not work in the easiest way. It works at every point in a way that provokes criticism and slows down progress. And the point is precisely to guard against the excess of power. And that is not done, in my opinion, by eliminating progress in the techniques but by guaranteeing the adversary functions within our government so our government should limit itself.

I remarked that Teller himself makes no bones about being highly critical of those who disagree with him. Was he now saying that he really had no problem with them?

> Of course. I become indignant, not when they win, but when they win easily. When it becomes obvious and clear to everybody that nuclear explosives are an evil. When everybody is afraid of radioactivity, when there is an exaggerated fear of the ozone depletion. Then I see, not an argument which I consider wrong, but an argument that claims to be obvious.

I noted that we had again come around to the problem of the general public being able to understand technological issues. He agreed. "Precisely. And furthermore, I have to admit that the job to explain science, including relativity and quantum mechanics, to everybody is very hard and therefore its neglect, although terrible and deplorable, is understandable."

At this point we had gone over the one hour allotted for the meeting, and Teller's next appointment was waiting in the outer office. I asked if he would grant me a follow-up interview the next day and arranged to call him in the morning. As I prepared to leave, Teller continued to discuss his final point, saying that there had once before been a scientific revolution with claims seemingly as absurd as those of the twentieth century. He was referring to the Copernican revolution and the idea that the earth was not stationary but moving. He suggested I read Arthur Koestler's book on the subject, *The Sleepwalkers.*

> Now that [the heliocentric worldview] in the course of time, and not without opposition from the Church at that time, was finally assimilated. So that now, somehow a child can understand that the earth is moving. I say it will be necessary to understand quantum mechanics and relativity in the same primitive manner. And that, in fact, has not happened.

I turned off the tape recorder and was shaking Teller's hand when he said he wanted to ask me something. What, he wondered, was my impression of Hans Bethe? His question surprised me. Unsure of what he wanted to know, I simply said that I liked Bethe very much and quickly added, "He is a very intelligent man, as I can see you are, Dr. Teller." At that moment Teller's colleague entered the room bringing some cookies. I left and was in the outer office chatting with Patricia French when I heard Teller call out, "Give her a chocolate chip cookie, she deserves it!"

On the trip back to my hotel I munched on the cookie and thought about Edward Teller. In some ways I found the controversial "father of the hydrogen bomb" as I had expected him to be—intense, uncompromising, opinionated. However, I had not anticipated that once we got into the discussion he would be charming and animated and possess a good sense of humor. At one point during the interview, I began to use the word *irregardless,* something I never do. When I told Teller of my imminent faux pas he laughed and replied, "Irregardless. Irregardless sounds to me a little more Hungarian!"

I had imagined that the famous hawk would be ponderous and obvious in his thinking. But, during the course of our conversation, I became aware of the subtlety and sharpness of his mind—he was, after all, in his late eighties. I saw something of the romantic in him, certainly something of the idealist. But I was most struck by the sense of him as someone fascinated by his own thinking process. Sometimes Teller seemed to be not so much talking to me as caught up in his rapid internal mental connections. After the interview I recorded only one note in my journal: "Edward Teller: A mind delighting in its own quickness."

After our first encounter, many of my questions remained unanswered. In July 1945 Teller had written to Leo Szilard, "First of all let me say that I have no hope of clearing my conscience. The things we are working on are so terrible that no amount of protesting or fiddling with politics will save our souls."[10] What had he meant? And how could the democratic process in which he so fiercely believed possibly work when the very social structure of the nuclear weapons culture was built on secrecy? How could the public influence decision makers without adequate understanding of the scientific or technological complexity involved in such decisions? Finally, I had lingering questions about the connections he made between science's search for the truths of the natural world and the technological applications of that knowledge. It is one thing to understand the physical laws necessary to produce a nuclear weapon, quite another to make the decision to bring that weapon into being—into social, political, and historical reality. I was determined to focus on these questions at our meeting the next day.

In the morning I called Teller to confirm our meeting and to ask his permission to bring my husband, Joseph, to the interview, as we were en route to the airport. He assented, and late in the morning we arrived at his Palo Alto home. Approaching the house, we heard strains of a Beethoven piano sonata. A medical oxygen sign was posted on the front door, and when the housekeeper greeted us, we saw the physicist rising from the piano. I introduced Joseph to him, and as he led us to the dimly lit living room, I noticed a table covered with Teller's medals, which he told me his wife collects and displays.

We took our seats, and I began to set up my recording equipment. After a few minutes, I heard a woman's voice and out of the corner of my eye saw a cord moving on the floor. A tiny, frail-looking lady using oxygen entered the room, and Teller introduced Joseph and me to Mrs. Teller. Mici Teller, who has been married to Edward for more than sixty years, greeted us and took a seat on the far side of the room directly facing her husband. During the next two hours, she listened silently, murmuring one or two questions as Teller spoke. She seemed to me like a pale guardian angel, protectively watching over her husband.

Although I was determined to ask some very specific questions, once again Teller took control of the interview from the outset. He began by telling me why I had come back. "Look, you want to know about my religion—what I believe in." Although this was not what I had intended to ask, I was intrigued and curious to hear what he had to say.

> And that I can tell you about very briefly, and then I can tell you about it in several hours if you want to. I believe in science. I have, therefore, strong feelings about people who disagree with me on science, and most of them are scientists. I believe you know, contrary to what is the general impression, that science is by no means finished, that science consists of surprises, that these surprises are of course, by definition, unpredictable.
>
> I literally grew up, you know, at the very end of the last magnificent period of science, which produced relativity and quantum mechanics. That was a period from 1905 to 1930, and I became a physicist in 1930. I came in 1928, I was a very young physicist. And then I saw the destruction of this by Hitler, and the very great changes in it, by overemphasizing applied science, in which I participated very strongly. I am by all means for the application of science, but I am not for the replacement of science by applied science.

Teller told me that when he was young, he had wanted to be a mathematician, not a scientist. He recalled in some detail his early fascination with geometry and his reading of Euler's text at age eleven. Although his father encouraged this interest, he did not support him in pursuing mathematics as a profession. The only career available for a mathematician was that of a university professor—something that, as a Jew in Hungary, Teller could never become.

> So, after some haggling, we compromised on chemistry, and then I studied chemistry for a couple of years, from '26 to '28. And by that time learned enough about mathematical physics to get interested, really interested in the exciting new things in quantum mechanics. And that is

where I got to Heisenberg in Leipzig. I was then twenty years old, I got my Ph.D. when I was twenty-two, stayed in Leipzig for another year, got to Göttingen, worked there as assistant professor for a couple of years, then Hitler came and I went. And within three years I landed as a professor at [George] Washington University.

When, in the context of our discussion, I mentioned my own father, Teller said it seemed that my father had "had something of a bad conscience about the bomb." I replied "Yes," and he added,

> And I am trying to tell you that he was wrong. I mean what I'm trying to tell you is that he wasn't and shouldn't be responsible. That he was responsible for doing good work. Look, I tell you, having a bad conscience about that is, unfortunately, extremely fashionable among scientists and it has something to do with this having too high an opinion of oneself. I don't think we are all that important.

I appreciated his telling me that my father's work was of value. Yet this was not the first or last time I would hear negative remarks about scientists who experienced some personal conflict about the use of the atomic bombs. And, as usual, I felt a kind of anger begin to well up in me—I needed to defend or at least explain my late father. I told Teller that my father had not spent his life beating his chest, far from it. However, his Polish-born mother was from a family of rabbis and had taught him strong humanitarian values. Thus he had faced certain ethical issues within himself. And, I added, perhaps my father agreed with Teller that something like a demonstration should have been tried, to avoid the atomic bombings of the Japanese. He quickly answered,

> Look, I am not saying at all that it was wrong to bomb the Japanese, I'm saying very definitely and very loudly that *I don't know* whether it was right or wrong. And that I don't have the real instruments by which I could know. What I'm trying to tell you is, all right, I heard the arguments, but to hear arguments, and you know I read that horrible book by Alperovitz, for the simple reason that I do want to have the arguments from all sides. But to have the arguments is very different from

knowing how it happened, when it happened, knowing the details, know-ing the inflections of voices at that time. And that was not my business.

So we returned to Teller's "weak and strong regrets." I asked if I had understood correctly that his strong regret was that when Fermi asked him directly, he did not think about a possible alternative to the even-tual use of the atomic bombs. He replied, "Exactly." However, this re-gret raised a question in my mind about his actual reply to Szilard, which seemed to indicate that he was already resigned to the situation as it stood. I told Teller that I had a question about his July 2, 1945, let-ter, and he asked, "And what the devil did I say? I may not agree with myself." I read a passage from the opening of his response to Szilard: "First of all let me say that I have no hope of clearing my conscience. The things we are working on are so terrible that no amount of protest-ing or fiddling with politics will save our souls." Teller explained, "Look, I am talking to [Szilard] who did have that position. And I was putting myself into his shoes. So by starting that way, I was simply starting it by how he was looking at it. All right?" But what about the closing, in which again he seemed to be saying there was nothing to be done? I read from the letter, "I should like to have the advice of all of you whether you think it is a crime to continue to work. But I feel that I should do the wrong thing if I tried to say how to tie the little toe of the ghost to the bottle from which we just helped it to escape." He replied,

> Well, look, by that time the question of the hydrogen bomb was already very real to me. And clearly we were heading into a period where I was considered a criminal for working on the hydrogen bomb. And it was so unanimous that a very good friend, who was and remained my friend, like Fermi, refused to work on it, and he was much too wise to tell me not to work on it, but he implied it. That [the closing] was simply a ref-erence to continuing to work, which already at that time was in the air, we should stop. And I was the one who protested against that.

It seemed to me that if Teller had been able to put himself in Szi-lard's shoes regarding having a bad conscience, then it would follow that he could at least comprehend the general public's deep misgivings about nuclear weapons. He responded by telling me more about his friendship with his countryman. "Szilard did tend to feel that way, but he was not very strong on it. Szilard was very strong on seeing good sides in Communism, in the Soviets. With Szilard I differed less on the hydrogen bomb and more on Stalin." Teller pursued the issue of the hydrogen bomb by telling me that after the war most of his colleagues went back to the practice of pure science.

> And for them, or most of them, the point that it would be wrong to continue [research on the thermonuclear weapon] turned into a motivation, which did not take place in my case. What you read here [the letter to Szilard] is the very beginning of that controversy, when it was not yet spread out.

Surprised by what he was saying, I asked Teller if I had understood correctly that the closing of his letter to Szilard—"I should like to have the advice of all of you whether you think it is a crime to continue to work"— was a reference to his continuing work on the H-bomb. In the same letter he had written that the only hope was to make the weapon known to people, that this might help to convince the world that the next war would be fatal: "For this purpose actual combat use might even be the best thing."[11] Therefore, I had assumed he meant work on the A-bomb. Teller answered,

> That would be the immediate application, yes. That was a natural continuation on which I was actually, even then, working. And where I was simply ahead of everybody else. Not because it was so difficult, but because nobody else wanted it. Now look, I have to add something that was also important. Getting away from Hungary and from Europe, apart from my closest family, which at that time consisted exclusively of Mici, I had absolutely nobody except the scientists. And now the great majority of them came up with value judgments of the work that were in complete variance, in complete variance, with what I was, in the main, trying to do.

My husband, Joseph, interjected that he must have felt very isolated from his colleagues. Teller replied,

> Of course I was. Of course. Excuse me, I was not isolated, I was a criminal, I was worse than isolated. Look, many people, including [Richard]

Rhodes, you know, give a description of me that my whole motivation was the hydrogen bomb. The simple fact is that I did not want to work on it. I wanted *others* to work on it. And I wanted to work on it only then when I saw that nobody else was willing to.[12]

With the success of our earlier work, with the reaction of the bulk of the scientific community to that, I saw the continuation of the work on the hydrogen bomb, not [only] in danger, but come to an end. And I was just very clearly convinced that that was wrong and that I should do something about it.

I asked why it was wrong.

Now let me, before I try to answer your question, tell you that I was in a unique position for a point that you know but that you may not realize at the moment. And that is secrecy, all right? Except for secrecy, this would have been an argument that could have been made and then left alone. If it's right, there will be people who worry about it. But the number of people who even could know about it were restricted. I talked with every one of them about it and I was not figuratively, but literally, alone. In other words, a man like [theoretical physicist Emil] Konopinski who worked on this [the thermonuclear bomb] with me from the beginning, you know, would listen, but he did not have the means or the strong interest to work on it if he was alone, you know? If I stopped on it, the thing would have stopped dead.

And what did he imagine the consequence would have been if that had happened?

I imagined that, and I now claim I would have been right, I was right. Had that happened, then Russia would be today Communist, and, I'm afraid, so would the United States. There was no question of holding back on the part of the Soviet Union. We know that quite independently from us they were developing nuclear weapons. They were behind us for only a few years. Had they gained an immense reinforcement and being in a position, unlike the Nazis, a considerable fraction of the population of the United States thought, well, all the story about Communism may be right. The cards were stacked very much in their favor.

Did he believe that, unchallenged, the Soviets would have developed the bomb and used it to conquer us?

> I don't have any doubt, I'm absolutely certain, that they would have developed the bomb. I have little doubt that they would have implicitly or explicitly threatened to use it. Whether they would have used it, I don't know. Look, one of the points about Soviet policy on which I'm talking to you, hopefully as an equal, but possibly, probably you know more about it than I do, one of the virtues of the Soviets was that they were patient. They were very much convinced that they were right, and that they would win in the long run. To what extent their power, together with patience, would have sufficed and through what intermediate steps, first influencing Europe and only then America, or whatever—I am certainly not good enough to invent for you a whole way of history, absolutely not. But it was a very powerful factor.

I asked Teller if he meant that his dedication to continued work on the thermonuclear weapon was in response to the particular situation with the Soviets. Or did he think it should have gone forward in any case?

> I had two independent and strong reasons. In principle, either of them would have sufficed. I did not want something to be stopped that was new, we had to find out. And I also did not want to stop when dangerous people were getting ahead of us. These were two entirely different things. And the circumstance that Hungary had been Communist when I was eleven years old for four months *did not*, as my opponents very clearly picture it, make me an anti-Communist. It did make me very much interested in Communism, and it did result in the point that when I got out to Germany, even then eighteen years old, for the following few years in Germany and partly in America, I got much more interested than the average Westerner in what was happening in Soviet Union. And in that course I became a dedicated anti-Communist, although not more so than I was anti-Nazi—if at all, less so.

I followed up on his first point. Was he arguing that the work should be pursued because it was new? He responded, "Right, that was, I would say, an independent half." But I wondered if a separation could be made between the theory and its applications—he seemed to be arguing that making the weapon was necessary.

> Look, excuse me, by working on the atomic weapon I learned that nothing will suffice, unless you actually do it. Look, here is a really eminent physicist, a very close friend of mine, Eugene Wigner, all right? Who in '42 and the beginning of '43 told me, with great emphasis and sympathy, "Don't go to Los Alamos, we know everything about it, there is nothing more to be done." And I get to Los Alamos, difficulty after difficulty, it is *not* done until you actually have done it.

Acknowledging that he might disagree with me, I said that it seemed we were discussing an intertwining of scientific knowledge and a particular application. A more dangerous weapon than had ever existed was being brought into being. Was that not a problem?

> Listen, there was scientific interest, and there was a political consequence. You are using the language that says the political consequence was negative, because it was a dangerous one. I say it was positive, because it served the stability of democracies as against the Soviet Union. What you consider, and most people consider, as a danger, I considered then and consider now as an advantage.

I told Teller that I accepted his point, but it still meant that both arguments are intertwined and that, in such a case, the scientific cannot be separated from the political. He answered, "I don't separate them out, but they were clearly acting in the same direction in my mind."

Teller insisted that the United States must continue technological development in order to maintain world stability. This raised another question in my mind. What were his present views on international relations as compared to those in his 1947 article "The Atomic Scientists Have Two Responsibilities"? Fifty years earlier, he had asserted that he and his colleagues had "two clear-cut duties: to work on atomic energy under our present administration and to work for a world government which alone can give us freedom and peace." The article concluded, "It seems difficult to take on these responsibilities. To take on less, I believe, is impossible."[13] Teller explained,

That I believed in at that time. And I still believe in it. Not in a sense that it will be effective tomorrow, but that we have to make a beginning of it. No, what I wanted to say is, that at the moment we are making negative progress in that direction. Because it would be important to begin to cooperate with the Russians.

When I asked if an agreement to stop nuclear testing might be a way of cooperating with the Russians, he responded, "No, *not* stopping. No, no. Look, stopping nuclear testing will not work. Stopping nuclear testing will contribute to proliferation and secret proliferation and that will give rise to tensions." Did he mean that with dangerous leaders in the world, the United States had to stay technologically strong while still engaging the Russians in a nonaggressive way?

Precisely. Use the complete development of everything possible for peace and for war and, of course, where there are peaceful applications, that I want to emphasize. Let me give you a very vivid example. I have been in Russia now twice. The first time in and near Moscow, the other time in the southern Caucasus, in the Russian-type Livermore—their second lab [Chelyabinsk-70]. What we discussed there was cooperation in using nuclear explosives to prevent collision from a big asteroid. You know the last time it happened was in 1908 when an asteroid about one hundred, one hundred fifty feet in diameter fell on Siberia. It may have killed one or two people, possibly no one. But it laid a forest flat for one thousand square miles. Now we estimate that that will happen once in every few hundred years. If it happens over New York, ten million people are dead.

The likelihood of that happening is very small, because the event is rare, and that it should happen in the wrong place seems less likely. To our knowledge the last big event occurred sixty-five million years ago. You know, that is the Alvarez asteroid. That was about ten miles in diameter and that exterminated something like two-thirds of the species on earth.

He prefaced his further explanation by telling me that the late Carl Sagan, then still living, was a strong opponent of his ideas about how to deflect asteroids because such technology had the potential to be misapplied.

Now, that they should then really hit us is very unlikely. I claim that these come close, we observe them, we know them, when they are past us and the danger from them is completely passed, then we send out the deflection apparatus and exercise it. Without this experience we are not going to stop one probably. Carl Sagan says, "Fine, fine, but what about somebody misusing that, somebody using it in such a way that asteroids should collide with us?" Now, to consider nuclear explosives as evil and try to take any application of it as wrong, no matter how improbable, no matter how clumsy, all right? Do I have to continue?

I told Teller that I hoped he would not like me any less for telling him that I believe we have to think about the potential danger of weapons that are so much more powerful than picking up stones and throwing them at an enemy. I began to say that misuse had to be considered whether or not it ultimately resulted in stopping development. Teller cut in, "Listen, listen, number one, of course it has to be considered. I like you less for only one reason. For the reason that you imagine that I don't know that." I immediately retorted that I was not imagining anything but simply trying to make myself clear to him in the conversation. He went on answering my point.

Well, of course, you cannot have a powerful instrument without looking at all its consequences. But to look only at the negative consequences is even worse. Look, I have already told you, but I want to repeat it. The whole history of destructive technology shows that the potential to do damage was, in any practical sense, unlimited from the beginning. That limitation was always from the side of the intention, and not from the side of capability. The past was when they came to the point that one side tried to exterminate the others, they had no difficulty.

If you get convinced that you can't survive, except by killing your opponents, and killing all your opponents, indiscriminately, you can do it. And we are not going to do it with nuclear weapons by mistake. The question is only the intention. Because the capability has been there for five thousand years.

He had criticized those who see all nuclear technology as evil. I wondered what he thought about people holding a negative view of him because of his advocacy of the development of such technology.

Well, look, they hold it in a more or less personalized manner. But the important thing is not that this view exists. The important thing is that it is so generally believed: atomic weapons are terrible, no matter how you look at them. You know I'm quoting. This is an incredibly stupid statement!

You know, you are asking, don't I agree that you have to watch out for the negative effects? Well, of course, yes, of course! Who ever thought of doing it without thinking of that? I am simply protesting against a statement that evil things dominate no matter how you look at it.

Teller continued to discuss the ways in which he believed science and the tools it creates are misunderstood. After a while, I heard the housekeeper preparing lunch in the kitchen and realized that it was time to bring the interview to a close. I asked Teller, considering the incorrect views that he believed existed, how he wanted to be remembered and how he wanted his work to be remembered. "I will give you a very concrete answer. First of all, let me tell you that your closing question is not original. I've been asked that before." I laughed, apologizing for my unintentional lack of originality, but replied that I still thought it was a good question. He told me that when he was first asked the question it came as a surprise, and unprepared, he had answered, "I don't care." I asked if he really did not care, and he admitted,

It's a lie, I do care. The truth is, I don't care very much. I say the issues we are talking about are so big that it's ridiculous to be overly influenced, "All right, I'm going to be remembered this way, I am going to be remembered that way." I am writing my memoirs, and that I do because I want to be remembered. And I'm not worried about, I'm not too excited. I am excited about what will happen. Let me tell you.

Then, gesturing, Teller asked, "See that thing there on the wall, above the key? You know what that is?" I stood and walked to where he was pointing and examined a plaque hanging on the wall. When I observed that it was something from Hungary, he responded,

It is very much something from Hungary. I hadn't visited there for half a century. And then when the Russians got out, I visited there. And I am saturated with the positive response to what I have done. I am very

much satisfied. Because in Hungary, it is recognized that I contributed
to it that the Russians are no longer there. And that's enough for me.
Whatever will be said about me after I'm dead is not important as compared to the point that the Hungarian government and Hungarians, in a
general way—independent of their Hungarian political position—appreciate what I have done. I am now not worried about how I will be
remembered, I'm now worried about what will happen. Whether the
present trend, the present antiscientific trend, the present fears of whatever is new, will cripple the United States and thereby eliminate the
huge stabilizing factor in world politics.

Even as our time ran out, my questions remained. Teller said that he
opposed secrecy and was in favor of scientific openness. But, I asked,
wasn't the whole environment in which he had attained his stature
built on secrecy? Hadn't Oppenheimer's 1954 security clearance case
been constructed on the question of the right to know secrets? Teller
told me that he had not opposed Oppenheimer because of his stance on
the hydrogen bomb, and I wanted to hear more about this controversial
subject. I knew that Teller had testified against his colleague. He recommended that I read a series of articles in the *Bulletin of the Atomic Scientists* written by Einstein, Oppenheimer, and Teller himself during the
early days of the hydrogen bomb controversy. As I listened, it occurred
to me that Teller was engaged in arguments I could not hear, with
people I could not see, answering questions I had not asked. He seemed
not to comprehend why so many of his scientific peers had so fundamentally disagreed with him. At that moment I believed he really did
not understand.

Joseph and I went to Mrs. Teller and thanked her for having us in her
home. Joseph admired a collection of Imari porcelain on the mantel. I
noticed some Southwest prints on the wall and seeing their date, 1944,
wondered if the Tellers had brought them from Los Alamos. As they
walked to the door, Joseph told Edward Teller that he did not consider
him a criminal. With a sigh, the physicist replied, "I did not intend to be."

✳ Martyrs to History?

I interviewed Edward Teller in December 1995, and the following January Hans Bethe was again at Caltech. I made arrangements to see him there. Three days before our appointment he called to suggest that I read Robert Maddox's book about the bomb and the end of the war. Bethe had been interested to discover that the historian's views about the decision to use the bomb were so close to his own. I obtained the book and read it the night before I drove to Caltech. Once again Bethe was sharing a small office with Gerald Brown, who, we realized, had known my father and had been to our family home in Brookhaven. We all chatted for a few minutes, and then Brown went off to the library so that Bethe and I could talk. We started with a discussion of Maddox's arguments in favor of the decision to drop the bomb, and Bethe again emphasized his conviction that the bomb saved lives, American and Japanese.[1]

As in the past, Bethe had outlined the topics that he wanted to discuss with me. This time he showed me a sheet of paper with four points, written in his careful, shaky hand. The first three items were talking points for our continuing dialogue about the decision to drop the bomb and Bethe's unanswered historical questions. The fourth was about the future.

Hiroshima was the right decision

Japan not ready to surrender

Why did Truman want the Russians?

Hiroshima cannot be repeated

Bethe's first and last points taken together seem to encapsulate his views of the atomic bomb's historical meaning. He told me,

Hiroshima cannot be repeated. In fact the whole Second World War cannot be repeated. One could not stage D day today. It would be noticed, and with long-range missiles those ships would have been destroyed. And Hiroshima, of course, for completely different reasons. This is just so horrible, and if it were now done with hundreds of bombs instead of two—I wanted very much to emphasize that Hiroshima cannot be repeated.

I asked Bethe to explain in more detail what he meant.

You can no longer use atomic bombs for saving lives. Hiroshima saved lives, lots of them, lots of Japanese and many Americans. If there were a nuclear war today, it would be a destruction of both countries, so in that sense it cannot be repeated. But I think the realization that it cannot and must not be repeated was very much facilitated by Hiroshima. If we hadn't had these two atomic bombings, people would not have realized what a terrible thing this is.

I noted that this view went against Edward Teller's argument that a demonstration would have been better. Bethe said he disagreed with Teller and added, "I think you had to see the center of Hiroshima leveled—completely destroyed."

I told Bethe that this was a difficult concept for me to accept. He seemed to be saying that the people of Hiroshima and Nagasaki were historical martyrs—that they had to die in the larger scheme of things, so that the world could know how terrible the nuclear bombs were. And yet he did not conclude that the horror of the weapon meant that he and his colleagues should not have developed it in the first place. He responded,

Correct. Well, of course, I have a number of reasons for defending Hiroshima, number one still is that it saved millions of Japanese, and probably close to half a million of Americans. And an argument against Hiroshima is that it brought nuclear weapons into the world, but I say that this had to be done. They were being developed by the Russians as well as by us. And it was important that both the Russians and we knew what it meant; and that the world knew.

Bethe paused and then added, "And so, in a way, the victims of Hiroshima died so that other people could live. It is unhappy, but that is the way it is. And there was no way of preventing the atomic bomb from being invented, both by the United States and by Russia."

While I understand that Bethe believes this, I struggle with his argument. First I consider the claim that more people would have been killed if the fire bombings had continued without the atomic bombs. An April 1945 meeting to review possible targets for the world's first military use of the atomic bomb made clear the scope of the United States' strategic bombing of Japan: the 20th Air Force planned to drop one hundred thousand tons of bombs per month on Japan by the end of 1945. The minutes of the Target Committee meeting recorded,

> It should be remembered that in our selection of any target, the 20th Air Force is operating primarily to laying waste all the main Japanese cities, and that they do not propose to save some important primary target for us if it interferes with the operation of the war from their point of view. . . . The 20th Air Force is systematically bombing out the following cities with the prime purpose in mind of not leaving one stone lying on another:
>
> Tokyo, Yokohama, Nagoya, Osaka, Kyoto, Kobe, Yawata & Nagasaki.[2]

To reason that by killing more than 150,000 human beings with atomic bombs, we stopped *ourselves* from killing thousands more with "conventional" bombs reveals the madness of the situation.* Yet I realize that those were the forces at work at the end of the war.

Then I consider the argument that in an invasion the Japanese militarists, while knowing they had lost, could have sacrificed even more of their people than the bombs killed. Although we can never know, when I read about the debates in the Japanese cabinet, this does not seem to be an unrealistic argument. The ministers in Tokyo appear so strangely

---

* Shortly after the war, the United States Strategic Bombing Survey estimated the dead from the Hiroshima bomb at between 70,000 and 80,000; from the Nagasaki bomb, at more than 35,000. More recent studies place the numbers, including those who died from acute radiation exposure, at 130,000 in Hiroshima and between 60,000 and 70,000 in Nagasaki by November 1945. J. Samuel Walker, *Prompt and Utter Destruction: Truman and the Use of Atomic Bombs against Japan* (Chapel Hill: University of North Carolina Press, 1997), 77, 80.

out of touch with what had happened to their country and its people. Even after Hiroshima, Nagasaki, and the Soviet entry, three of them argued for continuing the war. Admiral Toyoda, chief of the Naval General Staff, stated, "We cannot say that final victory is certain but at the same time we do not believe that we will be positively defeated."[3] According to Robert Butow, such arguments were not without justification. Summarizing the position of the war faction, he wrote,

> The Japanese military still had several million men in arms, some
> planes, and an assortment of materials of war. Their suicide tactics
> had been a serious threat at times, and that was while the fighting
> had been far from home. The navy had organized special attack
> squadrons composed of *kamikaze* submarines, explosive-laden
> motorboats, and human torpedoes. Because of past performance, it
> was logical to place great hopes in their future application—espe-
> cially when their employment would be within sight of Japan's
> sacred shores. The expendability of the loyal and the brave,
> although regrettable, was still an important tactical factor.[4]

I know that the actual circumstances of the atomic bombings of the Japanese cities cannot be debated outside the context of the whole brutal war. Nevertheless, I remain deeply troubled by the notion that the citizens of Hiroshima and Nagasaki were destined to die because of the deadly application of the scientists' discovery.

I asked Bethe if he used the same logic with the hydrogen bomb. He said that it was not the same, because even though we were on bad terms with the Russians, we were not at war. "If both governments had been sensible, it needn't have been developed. We certainly were in good communication with the Russians." I then raised Edward Teller's question, "Can lack of knowledge ever contribute to stability or safety? The hydrogen bomb controversy marked the first time that a large group of scientists argued for remaining ignorant of technical possibilities."[5] Bethe replied, "Right. Correct statement. And, well, I would have been very happy if we had remained ignorant."

CHAPTER 3

Philip Morrison,
Witness to Atomic History

On December 6, 1945, four months after Hiroshima, twenty-nine-year-old physicist Philip Morrison testified before Senator Brien McMahon's Special Committee on Atomic Energy, created to investigate problems relating to the development, use, and control of atomic energy. Morrison was one of several atomic scientists called to testify. Los Alamos director Robert Oppenheimer and head of theory Hans Bethe had been before the senators the previous day. Later in the week the witnesses included Leo Szilard, Samuel Goudsmit, and John Simpson. Dutch-born physicist Goudsmit had led the Alsos mission, an intelligence campaign to find out the state of the German fission project and to dismantle it. Alsos entered Germany with the first Allied troops. Simpson was one of the younger American-born scientists at the Metallurgical Lab. At the time of the hearings, he was chairman of the executive committee of the Atomic Scientists of Chicago and a member of the Federation of Atomic Scientists. Simpson cofounded the *Bulletin of the Atomic Scientists* and throughout his long scientific career has remained dedicated to educating the public about nuclear issues.

The lawmakers also heard testimony from military personnel, including General Leslie Groves, commanding officer of the Manhattan Engineering District, and industry executives from companies associated with the project. The senators and their witnesses were attempting to comprehend the implications of the atomic bomb, then in its infancy, for the future of war, energy, and international relations.

Morrison, like many atomic scientists, was deeply concerned about the postwar meaning of the bomb long before the senators heard the faintest rumblings of the weapon's distant thunder. His connections to the nuclear weapon were forged several years before the project itself had been conceived. During the late 1930s, he had been one of a circle of brilliant young theoretical physicists studying under Oppenheimer at the University of California's Berkeley campus. Morrison completed his doctoral work in 1940 and taught physics until late 1942, when he joined the Manhattan Project. He worked first at the Metallurgical Lab and then at Los Alamos. As a member of the Trinity project team, he witnessed the first atomic bomb test near Alamogordo, New Mexico.

After the successful test, Morrison accompanied the bomb components to Tinian, which lies fifteen hundred miles south of Tokyo and east of Manila in the Mariana Islands. "Fat Man," the bomb dropped on Nagasaki, was the second version of the one tested at Trinity. It was a plutonium implosion bomb in which the active material is imploded by surrounding explosives. The device nicknamed "Little Boy" had not been tested before its use on Hiroshima. It was a uranium gun, in which the two bomb halves were shot against each other. The design was simpler than the implosion bomb but used the costlier U-235 fuel. On Tinian, Philip Morrison worked on the bombs' final assembly.

At the time of the Senate hearings, Morrison had been back in the United States only two months. He and his colleagues in the newly formed Federation of American Scientists were committed to educating the senators, and the public, about the difficulties that lay ahead in the nascent nuclear era. They understood that because of secrecy and the complexity of the bomb the public lacked technical information. And they passionately believed that the government needed to face the new challenges to age-old assumptions about weapons and their power: the atomic secret could not be kept, for it was nature's to reveal; there could be no Anglo-American monopoly; there was no defense against nuclear weapons; more powerful bombs could be made; there had to be international control.

Morrison's knowledge of the bomb was broad, both scientifically

*Figure 11. Philip Morrison,
Los Alamos, 1945.*

and socially. In addition to his participation at Chicago
and Los Alamos, at Alamogordo and Tinian, he had the benefit of one
experience none of his fellow witnesses that week shared. He had been
a member of the preliminary American party sent to Japan shortly after
the nuclear devastation of Hiroshima and Nagasaki. There he witnessed
what he named "nuclear war in embryo."[1]

On that winter day in 1945, he told the senators,

> The reporters and the photographers have made clear for us all
> the appearance of the war-damaged towns, especially of the cities
> destroyed by the first atomic bombs, Hiroshima and Nagasaki. But
> there is more to be learned from those scenes than the newspapers
> have yet been able to tell. It was my job to visit the damaged cities
> of Japan, to speak with the people there, and to assist in the carry-
> ing out of certain technical studies. . . . I am a nuclear physicist, not
> a specialist on this or that kind of damage; I wish I knew even less
> about damage than I do. It is my purpose to tell the committee as
> clearly as I can what the impressions of an American physicist are

when he views the ruins and talks to the survivors of the bombing
he and his coworkers spent so much time to make possible.

Morrison emphasized that the atomic bomb was more than a new
weapon; it was a revolution in war making, and therefore in history. It
was not the virtual destruction of great cities that was new. Rather, it
was the means by which this destruction had been accomplished. He il-
lustrated this change by first detailing the massive effort required to
launch the devastating fire-bombing missions over Tokyo, Osaka,
Kobe, and other Japanese cities during the final days of World War II:

> From the air this island [Tinian], smaller than Manhattan, looked
> like a giant aircraft carrier, its deck loaded with bombers. . . . Here
> were collected tens of thousands of specialists, trained in the opera-
> tion and repair of the delicate mechanisms which cram the body of
> the plane. In the harbor every day rode tankers laden with thou-
> sands of tons of aviation gasoline.
>
> . . . And all these gigantic preparations had a grand and terrible
> outcome. . . . Down the great runways would roll the huge planes,
> seeming to move slowly because of their size, but far outspeeding
> the occasional racing jeep. Once every 15 seconds another B-29
> would become air-borne. For an hour and a half this would con-
> tinue with precision and order. The sun would go below the sea,
> and the last planes could still be seen in the distance, with running
> lights still on. . . .
>
> Next day the reconnaissance photographs would come in. They
> showed a Japanese city, with whole square miles of it wrecked and
> torn by flame. The fire bombs dropped on wood and paper houses
> by the thousands of tons had done their work. A thousand B-29's,
> time and again, burned many square miles of a city in a single raid.

The case of the atomic bomb, Morrison argued, was something quite
different. A small group of scientists and technicians had come from
Los Alamos and converted a few Quonset huts into testing laboratories.
Within weeks of the scientists' arrival, three planes took off from a de-
serted airfield in the middle of the night. A few hours later one plane's

bomb destroyed a city more thoroughly and with less chance of resistance than a strike carried out by one thousand planes.

Reading the testimony, I clearly understood that the monumental effort that produced the devastation of the atomic bombs was not visible on Tinian the night the *Enola Gay* left for Hiroshima. The work did not take place on the airfields and in the harbors. It had already been accomplished in Chicago, Hanford, Oak Ridge, and Los Alamos. There some of the world's most gifted scientists along with their young apprentices had mastered nature's recently revealed laws. With technical ingenuity, they had packed their new knowledge into two small bombs that were released on the unsuspecting citizens of Hiroshima and Nagasaki.

Throughout his testimony, Morrison attempted to show the senators the far-reaching implications of what he argued constituted a historic transformation of war and its meaning. "I can imagine a thousand atomic bombs in an airport like Tinian's to send them off. But not even the United States could prepare a thousand Tinians with ordinary bombs. There are simply not enough people. Destruction has changed qualitatively with this new energy. War can now destroy not cities, but nations."

Morrison spoke these words fifty years ago. Since that time we have grown accustomed to the idea that nuclear weapons can destroy nations, even worlds. But most of us only understand it in the abstract. What this actually means for us, and what it meant to the life of Hiroshima or Nagasaki, is still not well comprehended. Today we are shocked by the destruction of small bombings in London, Belfast, Beirut, Jerusalem, and Oklahoma City. Morrison's half-century-old testimony remains crucial, historically and for the future, because he looked, without blinking, at the awful consequences of the weapon he and his colleagues had made.

Morrison told the senators about a lunch he had attended with the Japanese officials while in Hiroshima. Each man gave horrifying accounts of suffering. The American physicist spoke with the chief medical officer who lived about a mile from the bomb's point of impact

and had been pinned under his house for several days. He recounted the
Japanese dignitary's story.

> His assistant had been killed, and his assistant's assistant. Of 300
> registered physicians, more than 260 were unable to aid the injured.
> Of 2,400 nurses, orderlies, and trained first aid workers, more than
> 1,800 were made casualties in a single instant. It was the same
> everywhere. There were about 33 modern fire stations in
> Hiroshima. Twenty-six were useless after the blast, and three-quar-
> ters of the firemen killed or missing. . . . Not one hospital in the
> city was left in condition to shelter patients from the rain. The
> power and the telephone service were both out over the whole cen-
> tral region of the city. Debris filled the streets, and hundreds, even
> thousands of fires burned unchecked among the injured and the
> dead. No one was able to fight them. . . .
>
> With the doctors dead and the hospitals smashed, how treat a
> quarter of a million injured? . . .
>
> A Japanese official stood in the rubble and said to us: "All this
> from one bomb; it is unendurable." We learned what he meant.[2]

Morrison agreed that this was indeed unendurable. He explained to
the senators that it was impossible to tell when a bomber armed with
an atomic weapon was approaching—that people now had to live in
constant fear of death from a single plane that gives no warning that
they should seek shelter. And in later testimony before the same committee, Morrison admitted that the scientists had failed to comprehend
the full impact of their creation's instantaneous, near-total destruction.
"We did underestimate—and this was an error—the terrible effect on
casualties, and the great morale and psychological effect that came
from the saturation nature of the bomb."[3]

Morrison presented a detailed explanation of how and why the
atomic bomb worked in a way so different from any other weapon. Det-
onating a bomb in the middle of a city, he said, is like creating a small
piece of the sun. Those near the sun will be burned. He described the
explosion; the shock wave; the 500- to 1,000-mile-an-hour winds; the
flesh-burning radiant heat; the instantaneous burst of radiation from
the explosion; the radioactivity deposited on the ground.

One senator asked if the radiation from the explosion could result in third-degree burns, and Morrison explained that such exposure would result only in a slight burn. However, he added that many Japanese citizens who had escaped death from the tremendous heat of the explosion, from collapsing buildings or fires, were nonetheless doomed.

> They died from a further effect, the effects of radiumlike rays emitted in great number from the bomb at the instant of the explosion. This radiation affects the blood-forming tissues in the bone marrow, and the whole function of the blood is impaired. The blood does not coagulate, but oozes in many spots through the unbroken skin, and internally seeps into the cavities of the body. . . .
>
> The white corpuscles which fight infection disappear. Infection prospers and the patient dies, usually 2 or 3 weeks after the exposure. I am not a medical man, but like all nuclear physicists I have studied this disease a little. It is a hazard of our profession. With the atomic bomb, it became epidemic.

After hearing Morrison's clinical description of the grave dangers of radiation sickness, some of the senators questioned him about strategies for protecting people from radiation after an atomic attack. Morrison explained that the bomb at Trinity had been detonated close to the surface, resulting in significant radioactivity on the ground. This was not the case in Japan, where both bombs had been detonated at such a distance from the ground that the radioactivity spread over a wide area, resulting in a concentration that Morrison judged to be "negligible." The senators wanted to know what materials could be used to shield people from high concentrations of radioactivity, particularly in a major city, where essential functions must be kept going. When asked what sort of protective clothing could be worn, the physicist explained that effective protection would require fifty tons of lead. He emphasized that it is impossible to shield people from the radiation with the amount of material they can carry on their bodies.

The logical end point of Morrison's presentation was that the bomb had created a qualitative change in the nature of war. The accumulation of destructive power concentrated in a single weapon represented an essential shift. To illustrate what he meant by a qualitative change, he gave the example of the incremental heating of water, during which the liquid grows hotter and hotter until, at a certain point, it is transformed into something completely new—steam. This analysis brought a challenge from one of his questioners, who argued that the change was essentially quantitative—simply *more* of the same order of destruction—but Morrison insisted otherwise.

> By the piling on quantitatively you have produced a new mode of action which is in nature qualitative. . . .
> When you have a collection of atomic bombs, you have something different, then, which becomes a qualitative difference, because there you have a chance to destroy whole nations.

Morrison's reasoning led him to conclude, as had many of his colleagues on the project, that another war must be avoided. This would require the international control of atomic energy, and he strenuously argued for its necessity. The historic moment had provided a unique opportunity to create peace from the novelty and terror of the bomb. Morrison insisted that action in this direction had to be taken immediately, while the weapon was still young.

> I think that if we catch hold of this opportunity and soon begin control, you will find it to be a much easier situation to handle than if we sit around 1 year, 3 years, 7 years, 10 years, making no progress. Then, when it has come to a full-fledged realization, by that time there will be greater difficulty. It will be widespread; there will be money in it; there will be great industries, and the economy of great regions may be dependent on it. It will be sizable, and it will be great, and it will be difficult to make such changes as should be made, because . . . people will say that it is not practical.[4]

Looking back on Morrison's words from fifty years of history, it would seem that he possessed the power to see into the future. For today, in this millennial era, as he predicted it would, the reality of nuclear weapons has invaded many realms of the lifeworld. But Morrison did not have Cassandra's prophetic gifts—this I understood when we met. He applies to history and human affairs the scientist's long view of time and the nature of change. He looks for patterns in seemingly random events and for mutability in what appears to be stable. Through extrapolation and imagination he had predicted what the future would be were it lacking strong international control.

But he was like Cassandra in one way. His warnings were fated to go unheeded. Perhaps Apollo was angry at the scientists for having imitated the sun.

I met Philip Morrison in his office at the Massachusetts Institute of Technology on a muggy New England midsummer's day. After the war he had returned to academic life, first at Cornell and since 1965 at MIT, where he is Institute Professor Emeritus. In addition to his teaching, he has dedicated himself to promoting an understanding of science and arms control to a wide audience. He accomplishes this through books, articles, television programs, and, for thirty years, monthly book reviews on all aspects of science for *Scientific American*. While preparing for the interview, I developed a respect not only for the wide expanse of Morrison's ideas, from the scientific to the social, but also for the eloquence with which he expresses them.

I arrived at MIT after a considerable hike across Cambridge from my parking place on a crowded side street. I hurried down the institute's halls, the heels of my pumps clicking as students casually strode by in their Nikes and Reeboks. I found Professor Morrison sitting behind the desk in his large, airy, book-lined office. Suspended from the high ceiling was what seemed to me an unlikely creature, a crocodile, made of cloth. Morrison's wife, Phylis, later explained to me that she had crafted it for him—the crocodile being a medieval symbol for the alchemist.

Philip Morrison is a small, delicate-looking man with a youthful energy emanating from his eighty-year-old frame. Soon after our conversation began, I became aware of his expressive hands and large blue eyes, which, when he speaks, rest intently on the listener. He wore a beautiful old silver belt buckle that I learned was an artifact from his days at wartime Los Alamos. He told me that he bought the traditional squash blossom piece from an artist at the Santa Fe Indian School and has worn it nearly every day since.

As I set up my recording equipment, Morrison quietly noted that it was July 16, exactly fifty-one years after the first atomic bomb test in Alamogordo, New Mexico. Unlike the previous summer, when the country had been in an uproar over how to properly commemorate the fiftieth anniversary of the atomic bombings and the end of World War II, this day was passing into history with little more than a murmur. But, for people like Morrison, the memory would always be vivid. "The Trinity Test, the first test of a nuclear bomb, went off as planned, on July 16, 1945, leaving lifelong indelible memories. None is as vivid for me as that brief flash of heat on my face, sharp as noonday for a watcher 10 miles away in the cold desert predawn, while our own false sun rose on the earth and set again."[5]

One of the first things I wanted to know was whether Morrison had done any weapons work after the war. He replied that he had not, quickly adding that nobody would have asked him. Morrison had been a vocal student activist at Berkeley and a member of the Communist party from 1936 to 1942. In May 1953 he appeared before senators much less friendly than those in 1945—members of William Jenner's Senate Internal Security Subcommittee investigating left-wing subversion within higher education. Rather than the Fifth Amendment, Morrison took the "diminished Fifth," answering questions about himself but refusing to answer questions about others, including Jenner's extensive interrogation about his mentor and friend, Robert Oppenheimer.[6]

Morrison's later troubles notwithstanding, I wondered whether he had been clear about this decision not to continue weapons work at the end of the war. He replied, "I never had any doubt about that from the beginning. One nuclear war was enough. I couldn't imagine after the events unrolled that there would be another danger like that of the Nazis coming to us." His answer surprised me, because I expected him to say that it was the bombings of Japan, or perhaps the Trinity test, that had turned him away from weapons work. When I asked him to explain "from the beginning," he replied that he meant from the time he joined the project. From his earliest involvement in the bomb's creation, he had cast himself as witness to its world-changing nature. "I felt, it gradually developed, that I was in a good position to be a witness to many, many parts of the project. I thought that was a useful political position, historical position, to come in so I could write and talk about it and so on." Morrison's passion is to observe with a scientist's eye while simultaneously absorbing that which is before him. Then, using elegant prose to describe the events, he challenges his listeners and readers to confront the meaning of what he has seen—to examine their own thoughts, feelings, and assumptions.

He emphasized to me that the scientists were deeply afraid that the Germans would succeed in making an atomic weapon. Even as late as 1944 he was concerned, unreasonably as it turned out, that the Nazis were ahead in the race for the bomb. Morrison told me that it was not clear to him until November that the Germans posed no nuclear threat. But, simultaneously, many scientists were aware that their work had significance beyond the immediate.

When we had first spoken by phone, Morrison and I had discussed briefly the decision to use the bomb on Japanese cities. He had said that he agreed in general with Hans Bethe's views about the use of the bomb but also disagreed on some points. Bethe is convinced that the weapon was the least of many alternative evils and that history shows that it saved not only American but Japanese lives as well. Now, as we sat together, I asked Morrison if he could explain more fully how his views differed from Bethe's. After thinking for a moment, he replied,

> I don't think we know [that the bomb was a necessity]. I can't say he [Bethe] is wrong, but I don't agree that it is clear today. I felt that a warning with evidential support might also have worked. It would have been much better from the American moral position not to use it as a surprise, which was what was done. And I always spoke for that.

When I asked if by "warning" he meant a demonstration, as called for in the Franck Report, he clarified his position.

> Well, they [the Franck committee] intended a real demonstration, and I think that was not practical. But there is an intermediate step—a warning with movies and so on is quite a lot. You don't have to take a chance that it will fail, et cetera—all those arguments against it. Demonstration was extra melodramatic and wasn't really necessary. I thought even leaflet warning would have been enough, but they wouldn't do that either.

By "they," Morrison meant the members of the Target Committees, charged with selecting the targets for the atomic bombings. He recalled being one of the lab's technical people who served on two such committees when they came to Los Alamos. In this capacity he had an opportunity to participate in the planning for the actual use of the bomb. He remarked to me, "Essentially, I was given rather a hard time. I was excluded from having any real comment." I asked him if this was because he questioned the use of the bomb as finally decided, and he answered, "No, because I wasn't one of the same group." I later asked Morrison about the dilemma he had faced when it became clear his opinions were not being listened to in the meetings.

> I don't want to give the impression that I came away with a vast dilemma. I came away with the realization that we had little influence on what was going to happen. I did not understand enough about the issue to say anything, as I was so very well rejected by the one officer I talked to. I could see I could have no influence on a man like that whose thinking was of an entirely different nature. So, what was I to do?

I asked Morrison to reiterate what he said in the meeting.

> I said it would be very interesting to give a warning—and would be very good for the whole moral and political position. And he [the officer in question] said, "Well, you may say that, but I can't, because if we give a warning they'll follow us and shoot us down. It's very easy for you to say and it's not easy for me to accept." Which is as sure a put-down as you ever heard. It was more important because I didn't know much about war then. What I came to understand was that they lost bombers and bomber pilots on lesser grounds than that. They were quite prepared for sacrifices. So it was in some ways an honest, but not really a relevant, answer.[7]

I continued by asking Morrison if his position regarding using the bomb on Japan had been clear in late 1944, after the German defeat seemed assured. His reply reveals the labyrinth of issues into which the scientists were forced by their decision to work on a nuclear weapon.

> Well, I was not—I don't think I was in favor of dropping it on Japan, but I was in favor of continuing the obligation we had undertaken to the

United States, because too many people were dying in the war against Japan for us to assert an independent, antidemocratic process. That was wrong. And I also was caught up in the general view which I think Bohr expressed very clearly: that unless this was made public the world would be in very bad shape. What one imagined was that it [the bomb] would not be used because it would not be needed for decisive action and would remain secret and people would not know about it. And the powers then, of course, all know about it, and there would be kind of a hidden arms race, even worse than the one we had. That was the consequence that I was most afraid of.

Morrison told me that a public test seemed helpful because it could have countered secrecy and prevented a hidden arms race. "Whether that would have gone to the position of saying we should use it on Hiroshima," he said, "I don't know whether I would have gone that far or not. They didn't ask me. But I had an alternative proposition. They wouldn't do that either."

I wanted to ask Morrison a question of personal significance. Before his death, my father had told me of the communication between Chicago and Los Alamos scientists regarding alternatives to use of the bomb without warning on Japanese cities. When I mentioned this to Hans Bethe, he asserted that only a small circle at Los Alamos had known about the Met Lab scientists' demonstration idea. Bethe implied that, as a low-ranking electrical engineer, my father would not have been privy to conversations about the demonstration, unless he had been included because of his previous work at the Met Lab in Chicago. But my sense from my father's end-of-life talks with me was that he had discussed these questions at Los Alamos. I wanted Morrison's insight as one of the younger scientists. For although he had been senior to my father in the lab hierarchy, I thought he would know what had been whispered among the scientific workers' ranks. I asked him if ideas like the demonstration were being discussed to a great degree, and he replied, "Oh sure. Well—considering we were very busy—but to a considerable degree." In contrast to Bethe, he recalled that such issues were "pretty widely talked about."

I continued by telling him that, based on what I had learned from my parents, I have always thought that a demonstration or warning should have been attempted, that the surprise bombings of Japan had been wrong, but that my conversations with Hans Bethe caused me to open my mind to the other side of the argument. Morrison responded, alluding to the controversy surrounding the commemoration in 1995 of the bombings and the end of World War II. Much of the argument about how to represent this history centered on the planned exhibit at the Smithsonian Institution's National Air and Space Museum.

> Well, it was a very complex matter. I think it was wrong still. But the main point was not so much that the people had some harsh motive, it was more that they—you saw what happened last year at the Smithsonian—the people who saw the whole thing as a deliverance. And they were right about that from their point of view. But it was very narrow of them to consider that history ended on August 8 [1945], or whenever it was. You have to ask what happened for the next fifty years.

Morrison's point was that the veterans' views did not represent those of the public in general. After all, he asserted, today few people actually remember the events of the war.

> It's a very singular person who was present at the time. Those people all have to be seventy years old. So nobody forty years old remembers that. They don't see the war as very important in their direct experience. They might have read about it. But they see as important the day-by-day, the year-by-year problems of the cold war—the danger of nuclear war and the shelters and discussions and so on. So the view of most people is quite different than the view that asserted itself at that anniversary. Which was the view of the veterans, mostly of the air force, who saw the whole thing as an air force accomplishment.

What did he think would have happened if the exhibit had gone forward as originally planned?

> Then there would have been a balance. It said it [the use of the bomb] was a kind of a gate. Before there was this thing, and when you opened the door the war ended, and that was very good for the people. But what the open door opened upon was a long corridor of fear and

expense and danger which we've all been walking through. And that should not be forgotten either. And they had a considerable part of the exhibit to talk about that.

I remarked that there was also objection to the idea that the bomb was an act of violence against the Japanese people and culture. Some critics claimed that the proposed exhibit did not adequately represent Japanese aggression and atrocities.

They [the veterans] saw it as a deliverance. It just was a triumph of American power and justified by the Bataan death march—no use to worry about all that; that [the Japanese suffering] is another story. And as far as you go, that's true, but it's not a distinct story, it's the opening of the door. And they say if you're going to show how hard it was on the people of Hiroshima and Nagasaki, you have to talk about their death camps and the rape of Nanking and so on. But you can't really do that. You can't make the whole war now standing on this event. So I think the whole thing was curtailed strongly and it simply talked about it as the Americans would have seen it up until the day of the drop.

I believe the general public feels, as far as I can tell at least a large section of them, that it [the bomb] was a desperate catastrophe of unprecedented kind. And I think in some ways that's true, but only in some ways because of what it implied, not because of what it did at the time. What it did at the time was not as bad as the fire bombing [of the Japanese cities].

It was only in retrospect that you saw what happened thereafter, that you realized that it opened a *new kind* of warfare. And that was the defect of the American leadership, I always felt. That they didn't think of it as opening a new door. They thought it was closing a door—the end of the war.

Did Morrison, possessing knowledge shared by only a handful of Americans, conceive of himself as a witness to the "opening of the door" when he went to Japan after the surrender? Explaining his odyssey to the defeated island nation, he said,

When I went to Japan, yes, well, it all went by steps, you see. Then, of course, the other point, and I'm sure you would agree, is that the momentum of the institutions was very important. When you organize many people with tremendous passion to do something, they're going to do it. Even if the meaning of it has changed, it's very hard for them to see all that, especially all the way down the line. Especially when for many people it hasn't changed at all; it has only been intensified. And that was the thing. So I was asked and I wanted to go to Tinian because that was another step in the operation, that was getting close to the real event. And I was about the best person to do that in my group, so I volunteered. I was the person who had least family responsibilities and so on. So that was all right.

But having gotten there, and having had some background, as perhaps you remember in intelligence analysis, because for a couple of years I was one of the two or three people who worked as a sort of a part-time technical adviser to Groves's intelligence effort. So I knew something about the Japanese situation, not a great deal, but nobody knew a great deal. But there I was on the scene and General Groves wanted to send an early party to Japan. Partly because there was a lot of latent criticism about radioactivity, which was an interesting, genuine point, and partly just because something was happening he had to be involved in it.

What did he mean by "latent criticism"?

Of the fact that we induced radioactivity on the ground. And it was not yet the issue that it is today, but it was beginning to become an issue. So he [General Groves] wanted to go there and he wanted to be able to say that American, informed people had made the measurements and these were reliable measurements. He wanted somebody to go do that. And he was at the same time preparing a big, competent group of fifty or a hundred people who spoke Japanese, and who knew everything, to go and search the whole thing out in great detail—study the whole industry and the whole future, the casualties, everything, it was finally done. But, anticipating that, in the few days left to the end of the war, he knew he had a dozen people from Los Alamos in the Pacific who could be reconnaissance—an early party—an advance party, that's the way I came in. So, my general [Thomas Farrell] came and said, "Would you want to go and do this? Because the general [Groves] would like to send somebody." And several of us went.

As Philip Morrison told his story, I was struck by how young he had been at the time and remarked that it must have been an intense experience.

Well, it was. Of course the whole war was an intense, extraordinary experience. I was quite afraid. You know, the first night we were in Hiroshima we spent in a functioning Japanese army base—with the soldiers guarding outside, walking back and forth and challenging the passersby. They were still in uniform. The war was over, but they were not demobilized. No end of the war to them; they were still doing what they always did. And here were ten or so people, Americans, who had been responsible almost directly for destroying Hiroshima. We thought maybe they would not like that.

So that always impressed me as being a rather dangerous, volatile situation. But it was quite opposite, the Japanese feeling was absolutely opposite. Nobody wanted to talk about that, express that. Only one person I ever saw showed what I would have thought was the most likely response, a kind of resentment to opening this terrible attack on Japan. That was the radiologist from the University of Tokyo—Tsuzuki, Professor Tsuzuki. He came to me and he had these wonderful reports of his own, his thesis had been done in Philadelphia. And he did, strangely enough, whole-body radiation of dogs.

I was having trouble understanding Morrison's lowered voice and asked him to repeat his last words. He again said,

Whole-body radiation—x-radiation of dogs—to study the medical effects. And he showed me the papers and then he put his hand on my knee, and he said, "Now all I did was work with a few dozen dogs, but you Americans, you have carried out a human experiment, much more impressive." And, of course, it was a very ironic statement. So he was quite right about that.*

Morrison came home in late September 1945 and immediately joined other atomic scientists in their intense political activity—organizing,

---

* Some time after the interview, I located the abstract of the Japanese scientist's paper (Masao Tsuzuki, "Experimental Studies on the Biological Action of Hard Roentgen Rays," *American Journal of Roentgenology* 16, no. 2 [1926]: 134–150). His experiments had been done on rabbits, not dogs. When I wrote to Morrison about this in December 1998, he responded, "Such is memory; writing endures."

writing legislation, and testifying before Congress. He remarked to me that while he and his colleagues tried to produce a change in the situation at home, they influenced opinion but never won any battles. During that humid Cambridge afternoon, my conversation with Morrison ranged over a great breadth of subject matter. At the end of the interview, I asked if he found any lessons for the future embedded in this difficult history. He replied,

> Well, I have a very clear view on that, but it's not a very general one. It has to do with war. I think that it's probably so that contemporary science and technology and national war cannot very long coexist, or we'll just blow everything up. It's already being realized to some degree, but during the cold war it was not at all realized. The first person to my knowledge who said this in a striking way was not one of the usual people talking about it—John von Neumann—a *brilliant* person, very antagonistic to any reforming or policy change, very much part of the establishment and proud of that—actually quite influenced by the military.
>
> But, nevertheless, one of his last publications was a kind of a serious reflection on the situation in the world at the time. And he said the world has a very big problem which he doesn't know how to solve. And that is that the progress of technology is uniform—continues forever expanding—but the size of the earth is not. And we will soon cross the time when the technological effect of weapons is enough to destroy the living capabilities of earth. This is a very serious matter.[8]

On returning home and reflecting on my conversation with Morrison, I thought about the atomic scientists' concerns dating from the earliest days of the bomb project—the widening gap between technological development and society's moral development. The implication of documents like the Metallurgical Lab's Jeffries and Franck reports was that society's institutions would have to "catch up" with its technological capabilities. Yet Morrison was saying something more, which calls into question the validity of a basic assumption about our relationship to technology. It is not simply a matter of closing the so-called gap between technology and morality, if these terms are even correct or if such a thing could ever occur. His argument is that at the turn of the century we face fundamental questions about the very notion of unlimited progress and open-ended technological development.

Morrison's lifelong role as a witness to atomic history brought to my mind Niels Bohr's phrase "the great drama of existence, in which we are agents as well as observers."[9] Although Morrison had not been one of the Manhattan Project's division leaders, as a creative young physicist he had played an important part. I perceived in him an innate historical sense that gave compelling power to his recollections. I began to wonder if he had ever experienced personal tension between his participation in the Manhattan Project and his chosen role as witness and chronicler of these world-changing events.

I felt some trepidation as I asked him about this during a telephone follow-up to our initial meeting. After all, I have never experienced anything that comes close to those dangerous years and could not presume to comprehend what it had been like to live under the threat of Nazism. I have no way of knowing the kinds of choices I might have made had I been a young scientist then. However, I asked Morrison if he had felt any conflict, for example, when he was assembling the components of the atomic bombs on Tinian. He was, after all, both a witness and a significant contributor to this history. Morrison replied quickly and emphatically, "Of course. Well, I never hide that," and he concluded,

> Well, it tells you one simple thing. If you fight war you are going to do bad things, that's all. War is a bad thing; people who get involved in it have exculpatory reasons, they have moments that are different to point to, et cetera. But the fact of the matter is it's bad, and we can't live with it forever—especially not with high technology. That's the whole lesson as I understand it. And having goodwill doesn't make any real difference.

✳ Pacific Memories I

Just after Labor Day 1996, I set out for Boulder to interview David Hawkins. I was looking forward to speaking with him for many reasons. Hawkins and Robert Oppenheimer had met at the University of California, Berkeley, several years before General Leslie Groves, head of the Manhattan Engineering District, chose Oppenheimer to direct the bomb-building lab in the New Mexico wilderness. Then, in spring 1943, Oppenheimer summoned Hawkins to Los Alamos. As a philosopher, not a physicist, Hawkins was something of an outsider on the project. As Oppenheimer's friend, he was an insider. I wanted to know more about his views from this unique position, and I was curious to learn his insights into the complicated and controversial theoretical physicist.

I boarded the airplane at Santa Barbara, California, carrying a briefcase filled with reading material in preparation for the meeting. I sat on the aisle, hoping that at least one of the two seats next to me would remain empty so that I could put my books and papers beside me. While the plane filled, I began my work. I was reading one of Freeman Dyson's discussions of Oppenheimer's character. Dyson likened him to Michael Ransom, the protagonist of Auden and Isherwood's 1936 play, *The Ascent of F6*.

> [Ransom] is a mountain climber, known to his friends as M. F., a Hamlet-like figure compounded of arrogance, ambiguity and human tenderness. Over the years, as I came to know Oppenheimer better, I found many aspects of his personality foreshadowed in M. F. I came to think that *F6* was in some sense a true allegory of his life.
>
> The plot of *F6* is simple. M. F. is an intellectual polymath, expert in European literature and Eastern philosophy. The . . . accounts of his youthful exploits . . . resemble strikingly the stories of Oppen-

heimer's precocity and preciosity as a young man. As M. F. went to
the mountains for spiritual solace, Oppenheimer went to physics.
F6 is an unclimbed mountain of supreme beauty. . . .

At the end M. F. lies dead at the summit.[1]

I also had an Auden piece about the bomb and human consciousness:

Historical events since 1914 have destroyed the belief in inevitable
progress. It is clear that all knowledge, including knowledge of himself, and the power that knowledge confers, do not in any way affect
man's will; they merely increase his capacity to do what he wills, for
good or evil, and what he has actually chosen to do during the past
50 years suggests that, as theologians have always asserted, his will
is inclined to evil. . . .

Scientific knowledge about macroscopic or subatomic events
may enable us to perform many acts we could not perform before,
but it is still as inhabitants of this human world that we perform
them.[2]

Orange highlighter in hand, I marked the points, personal and philosophical, that I wanted to discuss with Hawkins. In the margins I scribbled questions and criticisms. I was looking forward to a solitary plane
trip—an intellectually stimulating journey of the mind in the company
of Dyson and Auden. It was to be something quite different.

September 5, 1996
Bus from Denver airport to David Hawkins interview in Boulder

*On the plane to Denver a tall, rugged-looking man who appears to be in his
seventies takes the window seat in my row. Once we are airborne, he places
his baseball cap on the seat between us. I look down at it and notice it is
embroidered with the words "USS Smith." The man asks me if I mind him
putting his cap there, and I reply, not if he doesn't mind sharing the seat. He
nods that he has no objection, and I place a file folder next to the cap and proceed with my work. The man looks out the window. When we reach cruising
altitude the flight attendants come by with sandwiches and drinks. I remove
my work from the tray table. As we eat, the man and I begin to chat. He asks
if I am a lawyer or a teacher. I reply that I am a writer. I ask if he lives in*

Santa Barbara. He tells me no, he has been visiting a son who does. As the conversation develops, I learn that he is an explosives expert who worked in strip mining for many years. He explains in detail the process of preparing a site for an explosion—how it sometimes takes three weeks of drilling before the explosives can be placed. He describes various kinds and mixes of explosives and how they work. I tell him that I have a connection to explosives as I am writing a book about the Manhattan Project. We talk about my research and about the bomb. I detect a subtle change in his demeanor—he seems curiously vulnerable and tentative. Then, as the Grand Canyon opens below us, he begins to tell me the story behind the baseball cap.

Mr. Hamilton [a pseudonym] *says that during World War II, he served in the navy in the Pacific. His ship, the destroyer USS* Smith, *took part in the Battle of Santa Cruz in the Solomon Islands. During the battle, a Japanese plane, possibly shot down by the* Smith *or another American ship, crashed on the* Smith*'s deck. Six minutes later the downed plane's unexploded torpedo blew up. Nearly sixty sailors were killed, and many were injured. His eyes fill as he tells me that his good friend, Lewis* [a pseudonym], *was among the dead. Mr. Hamilton was not at the crash site but was ordered to go forward to help out there. He describes a sickening scene of carnage and chaos. "I could only stay there for ten minutes."**

Unable to care for the many wounded on the destroyer, the crew strung a rope pulley between the Smith *and the battleship* South Dakota. *Wounded sailors were pulled across the Pacific expanse in their beds. Mr. Hamilton remembers that even after they were hit and badly damaged, the USS* Smith *had to protect the carrier* Enterprise. *"We were expendable," he says. The aircraft carrier was not. I remark that I can't imagine the bravery it took to go through what he and his shipmates experienced. He says, "I was not brave, I was afraid—it was survival."*

He tells me that after the atomic bombings of Hiroshima and Nagasaki he was sent to a Japanese port. There, armed with .45s, rifles, and a code of con-

---

* The battle for the Santa Cruz Islands took place on October 26, 1942, and was part of the Guadalcanal campaign. In the incident in question, a Japanese *Nakajima B5N2* ("Kate") torpedo bomber dived into the forecastle of the destroyer *Smith*. The various accounts I read differ in describing the plane crash as a Japanese suicide mission or as a result of being shot down by American *Wildcat* fire.

duct, he and his shipmates patrolled the city. They were instructed to "act civilized even if the enemy didn't" but were ordered to shoot to kill if any Japanese got out of line. But there was no trouble. He realized then that the Japanese were just ordinary people whose fate had been in the hands of the higher-ups in power.

He goes on to tell me that an amazing thing happened at the fifty-fourth reunion of the ship's survivors: the brother of a sailor killed on the Smith *attended in the hope of finding someone who had known his long-dead sibling. He brought a scrapbook containing wartime photographs. Mr. Hamilton was leafing through it when he saw his own fifty-four-years-ago eyes staring back from a faded picture. Then he saw a snapshot of the dead sailor and realized it was his lost buddy, Lewis. The owner of the scrapbook was Lewis's brother, who wept with the realization that after a half century his brother lived in the memory of a friend who had shared those terrible years and been near him at the end.*

When Mr. Hamilton finishes his story neither of us speaks for several minutes. My mind is on what he has told me of the battle. I cannot begin to comprehend how brutal it must have been. He breaks the silence: "One thing I will never forget—the smell of burning human flesh."

We are making our approach into the Denver airport. As the plane descends our encounter ends—we slowly emerge from the depths of his memories and once again breathe the air of everyday life. I have been leaning across the unoccupied seat listening, and now I sit up, straighten my clothes, and brush my hair from my face. We self-consciously reestablish social boundaries by shifting the conversation to our present situations. He tells me that he's now in the antique business. I reply that my late mother was an avid collector of antiques. As I describe details of carved table legs and curved chair backs, he speculates on what the pieces might be—American or English, eighteenth or nineteenth century. The plane lands and we begin our good-byes. He is changing planes for home, I am headed for Boulder by ground. Leaving the plane, we are separated.

Once in the terminal, I struggle with my carry-on bags and kneel to rearrange them. I glance up to see Mr. Hamilton stride by. Neither of us attempts further contact. I take the train to the main terminal. At baggage claim I pull my suitcases from the carousel and rush to the shuttle that will

take me to Boulder. I settle into my seat and the driver pulls away from the
curb. For a while I stare out at the prairie. Then I open my journal and begin
to write.

I looked up from my journal to watch the Denver airport sail away across the high Colorado flatlands like some massive, postmodern prairie schooner. As the sunlight faded, I wrote everything I could remember of my encounter with Mr. Hamilton. By the time the shuttle reached my Boulder hotel, night had fallen. I checked in, called David Hawkins to confirm our meeting on the following day, and phoned my husband to say I had arrived safely. I then ordered room service and took a quick shower. When my meal came, I switched on the TV and watched Gregory Peck in *Twelve O'Clock High*. In the 1949 film about American flyers in Britain, Peck plays a general trying to transform a group of frightened young men into a cohesive bomber unit flying daylight missions into France and Germany. Fresh from my meeting with Mr. Hamilton, I observed this war story with new eyes.

It occurred to me that much of my childhood awareness of World War II was shaped by such movies, the epics made in the forties, fifties, and sixties. Later I realized that I have never seen an unabashedly heroic war movie about the creation or use of the atomic bombs. In movies like *Twelve O'Clock High*, we in the audience accompany the young bombers on their flights and identify with them as they face the dangers of the German defenses. We cry for them as they are blown to bits, as they fight on with broken backs, surrounded by their dying buddies, their planes flaming and crashing. But we are usually not on the ground when their bombs hit—the story does not require it. It is enough to watch the brave warriors struggle and their generals lead and lose their men. It is enough to know that they are fighting against a formidable, often evil enemy.

Perhaps the difference is that the story of the atomic bombs cannot be divorced from its particular consequences—the instantaneous deaths of more than one hundred thousand civilians. The fire bombings of the Japanese cities killed equivalent numbers, but we can still focus on the flyers' heroism without looking closely at the results. Whether one believes that the atomic bomb attacks on Japan were right or wrong, the time between the act and the resultant deaths was so short, the space so concentrated, that film could not show the Allies' actions without immediately implying the horrific consequences for the Japanese victims. And, following on what Philip Morrison told me, the atomic bombs closed the door on a brutal war and simultaneously opened another onto the specter of nuclear annihilation. How could a movie depict all this and remain heroic?

CHAPTER 4

David Hawkins,
Chronicler of Los Alamos

I awoke to a steady rain, so instead of walking to my interview with David Hawkins as planned, I took a taxi to the address I had scribbled on a piece of paper. We stopped in front of one of several spacious old homes on a quiet, tree-shaded street. Hawkins, a slender man in his early eighties, with a kind, inquisitive face, greeted me at the door. I immediately noticed his beautiful silver belt buckle inlaid with turquoise. He told me that during the Los Alamos days he received it as a birthday gift from his lifelong friend, Philip Morrison, and Morrison's wife at that time, Emily, who was Hawkins's assistant during his writing of the Los Alamos project history. He said the buckle is so special to him that he refuses to wear anything that covers it up.

Hawkins is now Distinguished Professor Emeritus of Philosophy at the University of Colorado, where he has been for fifty years. His professional work has been devoted to teaching, research, and writing. In 1970, with his wife, Frances, he created the Mountain View Center, established to continue and widen the professional education of preschool and elementary school teachers. In 1981 Hawkins received a MacArthur Prize fellowship.

On entering the Hawkinses' home, I saw it was comfortably furnished, filled with art and books that I could see had been lovingly collected over many years. We made our introductions as I set up my tape recorder. After a few minutes, Frances Hawkins, who has dedicated her life to early childhood education, entered the room and introduced her-

*Figure 12. David Hawkins,
Boulder, 1996.*

self. She joined us for a while, taking a break from her
own work—reading the final proofs of her professional autobiography.
Seeing that I was not dressed for the cool weather, she brought me a
wool shawl while her husband prepared some tea. As she gently placed
it over my shoulders, I felt a pang—this was the kind of thing my
mother would have done.

As David Hawkins and I settled into conversation, the first subject
under discussion was Robert Oppenheimer. Hawkins explained to me
that he had come to know Oppenheimer at Berkeley in the late 1930s
when he had been a graduate student in philosophy and Oppenheimer, a
professor of physics. They were both activists on the political left, sharing "a common general political persuasion" that had to do with both
concerns about fascism and war in Europe and about wanting to be the
left wing of the New Deal. After completing his studies at Berkeley,
Hawkins taught for a year at Stanford and in 1941 returned to Berkeley to
teach. It was then that he and Frances got to know Robert and Kitty
Oppenheimer socially. The two men were drawn together intellectually

by Hawkins's philosopher's interest in quantum mechanics and Oppenheimer's study of philosophy. Hawkins commented to me,

> In some funny way I had a special place in Oppenheimer's regard. He
> knew my limitations, but I think he liked the fact that he could talk about
> the Bhagavad Gita with me—I don't know what it was. I never regard-
> ed myself as having the talents that he attributed to me particularly. But
> he certainly did.

Hawkins also laughingly told me that, Oppenheimer's respect notwithstanding, he was not spared the physicist's well-known sense of superiority. He recalled a visit to the Oppenheimer home during the Berkeley days.

> He could be quite arrogant, but also by turns quite humble. I was sitting
> next to a whole series of classics—Plato and Aristotle. And I said, mak-
> ing conversation, "I'm just reading the *Cratylus*." And he said, "I have
> looked at the Greeks, I consider the Indians deeper." Wow!

Hawkins said that the great Danish physicist Niels Bohr, Oppenheimer's mentor, was the only person whom Oppenheimer unre-
servedly regarded as his superior. He was devoted to Bohr both scientifically and philosophically. Although Bohr served as a scientific adviser
to the Manhattan Project, Oppenheimer paid closest attention to his
ideas on the future of the bomb—the potential for a nuclear arms race
and Bohr's plans to avoid it through postwar international control of
atomic energy. Bohr's interpretation of the larger meaning of the
deadly Los Alamos weapons work had great influence in crystallizing
Oppenheimer's thoughts and actions. And because of Oppenheimer's
relationship with the scientists working for him, Bohr's ideas had a
strong impact on them as well.

In June 1942 Oppenheimer had been appointed director of the
atomic bomb development project. Later that month, in Berkeley, he
had assembled a group of leading theoretical physicists, including Hans
Bethe, Felix Bloch, Robert Serber, and Edward Teller, to discuss bomb
theory and plan future work. Up until that time the Metallurgical Laboratory of the University of Chicago had overall responsibility for the
physics of bomb development. Concerned about compartmentalization

and other issues at the Met Lab, Oppenheimer was convinced that the bomb-building facility would have to be located elsewhere and suggested building a special laboratory. The idea appealed to the security-minded General Groves. In November 1942 the campus of the Los Alamos Ranch School was selected as the site for the enterprise, code named Project Y. It lay about forty miles northwest of Santa Fe, New Mexico, on a high mesa of the Pajarito Plateau. Oppenheimer and some staff arrived in March 1943. Hawkins recounted to me that a month later, while teaching at Berkeley, he received a phone call.

> I was walking down the hall one day and the secretary of the department called me and said, "You have a telephone call," and it was from Los Alamos. It was Robert's voice, and he said, "We need someone here right away, could you come?" That was a typical approach he would take. I think I had two or three days to wind up affairs.

I asked Hawkins if there was any question in his mind that he wanted to join the Los Alamos team.

> Not really. I wanted to be where my friends were—I wanted to be in on it. I didn't know for sure, but anybody who knew about the development of fission would have known that this was what it was. As Frances said, she was much more of a pacifist than I was. I wasn't thinking about the war in the Pacific nearly so much as I was thinking about the development of this technology and what it would mean. We were already onto that from talking with Phil Morrison, our close friend, and from Frances's brother, Leonard Pockman, who was a physicist—with all of these back-of-the-envelope calculations. So we were fully aware of the import of what was happening. And I think my reaction was just that I wanted to be in on it rather than outside of it—as a momentous thing that was going to happen.
>
> And Frances went to Los Alamos, as she said, somewhat reluctantly. She didn't voice her reluctance at that time, but I knew about it. So I went and was there within a few days. They even got me a plane reservation, which was very high priority stuff in those days. Frances and our eighteen-month-old daughter, Julie, came two months later.

I had read Hawkins's 1947 history of the Manhattan District's Los Alamos project. I therefore assumed Oppenheimer summoned him to

Site Y to serve as chronicler for the unfolding history. He explained to me that this was not the case. Oppenheimer needed an administrative assistant, in particular, to act as liaison between the lab's military and civilian populations—someone to deal with the tensions and ambiguous social roles created by the wartime laboratory's civilian/military collaboration. Hawkins recalled how confusing it was for officers who were accustomed to being at the highest levels of the social hierarchy to be on a military post where civilians occupied the top positions. There were, he remarked, obvious jealousies and misunderstandings. However, after he had been at the lab for a few months and had put procedures and routines in place, Hawkins was free to deal with jobs like getting draft deferments for young scientists, solving housing problems, and doing other administrative work and "odd jobs."

I wanted to learn more of Hawkins's views of his talented and controversial friend and boss. I knew that Oppenheimer was a great physicist and had skillfully managed and directed Los Alamos to unprecedented scientific achievement. Yet his character had remained rather one-dimensional in my mind. I confessed to Hawkins that my initial views of Oppenheimer had not been positive—I figured General Groves had found in him the perfect man to control. As gifted as Oppenheimer was, both as a scientist and as a leader, his ambition and left-leaning politics meant he was just so much brilliant putty in Groves's hands.

I had recently seen a documentary film in which Freeman Dyson characterized Oppenheimer as having made a Faustian bargain. Dyson maintained that Oppenheimer sold his soul to the devil for knowledge and power—that through his alliance with the army he had obtained the resources to do physics on the grand scale. Furthermore, Dyson observed, Oppenheimer saw himself as a philosopher king: "a man of wisdom who could get along with other men of wisdom who also had power."[1]

Yet, from Hans Bethe, I learned that Oppenheimer's particular genius had made the Manhattan Project a success. He possessed the ability to comprehend and synthesize wide-ranging theoretical issues and was the teacher of many talented and devoted physicists. I knew of the important schools he had developed at Berkeley and, postwar, at the Institute for Advanced Study at Princeton. I was finding that as I spoke to Manhattan Project participants, as I read their commentary and that of other scientists and scholars, the ghost of Robert Oppenheimer was emerging from the shadows.

Hawkins provided another perspective. He told me that Oppenheimer was a product of wealthy Jewish parents who had devoted a great deal of effort to the love and education of their two sons, Robert and Frank. He asserted that it was "not a negligible fact in Robert's background that he had been a victim of considerable anti-Semitism at Harvard and elsewhere." I realized there was an essential difference between Oppenheimer and Michael Ransom, the hero of Auden and Isherwood's *The Ascent of F6:* Ransom was a British golden boy, Oppenheimer a Jew.

Hawkins did not dispute that the elements of the Faustian bargain were there for Oppenheimer.

> He was obsessed with an understanding that this development [of the bomb] was in his view inevitable and that it was a change, to use his phrase, "a change in the nature of the world." Typically he saw that, and probably at the same time believed, and often correctly, that other people wouldn't see it. Therefore, he was early committed to this bargain partly because of a long-term political commitment which was not just of the political left. That phrase always stuck in my mind, "a change in the nature of the world." From now on it will be a different world. The potential control of energies of this magnitude makes it a different world.
>
> And he wanted to act in such a way that the world would understand this as deeply and as soon as possible. That, I think you could say, is a rationalization with which he cloaked the Faustian bargain. But I think it was quite genuine. I'm pretty sure from conversations with him that Oppenheimer really had the belief which would be, from an unfriendly critic, a rationalization of his position: that the bomb had to be used, because if it were not used in World War II it would be used in World War III. The world had to know about it and its full destructive character and not simply as a demonstration of an explosion, but as a weapon.

Well, obviously that could be, and in some measures certainly was, a rationalization of his position. But it has to be examined as an argument in its own right.

And I was well persuaded of it at the time, but in retrospect I suspect that the passage of events is much too complicated to make that kind of argument stick. Again, it's the lesser evil argument. The first lesser evil argument was the one about beating Germany to the production. The second one was, if we don't do this now something even worse will happen. But I think he was firmly convinced of that.

And remember that the Faustian compact, at least in Goethe's version, brought destruction but also the possibility of a new era: a fifty-year war that remained "cold," a nuclear pacifism creeping into high places.

In early 1945 Oppenheimer asked Hawkins to produce a technical, administrative, and policy-making history of the Los Alamos lab. Hawkins told me he was offered the job as "a sort of plum" after Oppenheimer's first choice, the eminent Stanford psychologist Ernest Hilgard, declined. Although he had misgivings about his competence because of his lack of historiographic training, he accepted the assignment. The resultant history is fascinating on several levels. It was first produced as a manuscript of the Los Alamos Laboratory, Manhattan District, U.S. Army Corps of Engineers. As the author, Hawkins had the privilege of deciding its security classification—Secret or Top Secret. Hawkins chose Secret—it did not, after all, contain a recipe for making an A-bomb. Regarding Oppenheimer's initial reaction to the work, Hawkins wrote,

> Oppenheimer remarked, rather darkly, that I was surely adept at "walking on eggs." He was given to dark sayings, and would carry the discussion no further. He must have meant that I had avoided or touched too lightly upon some of the conflicts and controversies of which he had, himself, been keenly aware. I did not take the remark to be altogether a compliment.[2]

The work was declassified in 1961, but between its writing and its declassification Hawkins had given up his security clearance and did not have access to the manuscript or his files. Then, in December 1950, he was investigated by the House Un-American Activities Committee because of his membership in the Communist party at Berkeley during the late 1930s and early 1940s. Coincidentally, he had left the party shortly before receiving Oppenheimer's offer from Los Alamos. Later this implied to some that he was part of a Communist infiltration of the Manhattan Project. Regarding Leslie Groves's attitude toward the politics of project members during the Los Alamos years, Hawkins remarked,

> General Groves probably believed that Oppenheimer was a Communist—he didn't give a damn. Anybody that would do this job was okay with General Groves. He wasn't of the McCarthy persuasion at all. He was in some sense unpolitical but sensible. And when I first got to Los Alamos it happened that his chief military intelligence person was at Los Alamos and wanted to see me. And he said very explicitly, "We don't care about people's past political affiliations, only their primary loyalty to the country."

Hawkins told me that when he was summoned to testify before the House, he feared that they were "gunning" for Oppenheimer, although at the time "they were still very cautious" about any public criticism of the eminent physicist. Both David and Frances Hawkins were called, and like their friend Philip Morrison did not take the Fifth Amendment but spoke only about themselves, refusing to testify about others. Hawkins told me that they were threatened with contempt of Congress, but it was the end of the second session of the 81st Congress, and the new Congress, busy with more important matters, did not cite them.

Thus it was not until fifteen years after completing his history of the Los Alamos project that Hawkins had his first opportunity to reread it. He has written of the disappointment he felt on doing so—the sense that he had recorded such extraordinary people and events in the restrained style of a military history. He told me several times that he could have caught more of the color, personality, and uncertainty of the era. "It could have been much more like a real history, but I wasn't up to that. I had no experience."

I told Hawkins that I had a different response to his chronicle. My strongest impressions on reading it had to do with the scope of the $2 billion project, the depth and breadth of the undertaking, ranging from payroll and employment contracts to abstract mathematics—from the mundane to the arcane. Reading it gave me a sense of the project as reality, not the myth of an elite group of scientists (some would say mad scientists) on the isolated mountain, unlocking the secrets of the universe and harnessing death. Of course, this was true on one level, but Hawkins dealt in detail with the day-to-day issues of a burgeoning government bureaucracy as well as the complexities of the scientific theory, research, and experimentation.

I began to see the Manhattan Project as the quintessential American enterprise, full of know-how and the pioneering "can do" spirit. It seemed so western in character. They built a city from the ground up atop a remote New Mexico mesa, not for railroads, cattle ranching, or gold mining, but for bomb making. It was a huge construction project with its connected logistical and labor issues. Housing, energy, water, and food had to be supplied. Payrolls for civilian and military workers had to be met. Scientific supplies were searched out and sent in from all over the country. And all of this had to be carried out under the deep cover of secrecy. Because the government's interest in scientific research had to be protected, patents were needed. The standard patent reports would reveal secret information, so a patent office was set up on the site. Hawkins's detailed description, which he considered technical and dry, produced in me an overwhelming sense of the magnitude of the project.

A complex institutional structure was developed at the laboratory, with divisions dedicated to theory, chemistry and metallurgy, research, ordnance, explosives, and the Trinity test. There were many groups in each division. Hans Bethe headed the Theoretical Division. Edward Teller's group, part of the so-called F Division, was dedicated to research on the "Super," which ultimately led to the development of the H-bomb. There were relationships with research laboratories nationwide. The business office was run by the University of California, and the lab's work was coordinated with that of Hanford, Oak Ridge, and Chicago.

As I read Hawkins's history I became aware that it was probably inevitable that the bombs, if completed before the end of the war, would be used. The institutional machinery created to achieve this goal was just too powerful, and it was moving inexorably toward an end that would be fulfilled. It seemed to me that it would have taken the most extraordinary circumstances—Japan's imminent and assured surrender—to avoid it. The technological structure required to make the idea of an atomic bomb a reality was too massive to respond to the sudden, subtle, often perilous changes that come at war's end. The effort to beat the Nazis to the bomb can be understood as a patriotic and unprecedented collaborative effort on the part of civilian scientists, industry, and the military. However, at the same time, the Manhattan Project can be seen as the military-industrial complex in embryo.

When his history was published, Hawkins looked back at the thirty-five years that had passed since its writing and described the work as "an early chapter in the biography of the bomb." He noted that later works addressed issues and developments to which he could not have had access in 1947. However, he was compelled to offer an observation that he had not originally made.

> There was one kind of development I must mention, a subjective one, which took place in parallel with the technical work of weapon development, and about which I surely knew at first hand. Everyone's life was being changed, changed radically I think, and irreversibly. Many of us were aware of those changes at the time, though I think even the most reflective of us were inadequately aware of them. I could not then have generalized about the nature of these changes, and I am not much more able to do so in retrospect. We all did know we were involved in something which would alter the nature of the world. We understood less, perhaps, the reflex effect upon ourselves. . . .
>
> A part of this reflex effect lay simply in the transformation of academic physicists, chemists, and mathematicians into creators of a radically new weaponry. More consciously in some cases, less in others, they were deeply affected by moral-political concerns about

the consequences, yet at the same time, flattered into self-consciousness by a sense of its overwhelming significance.[3]

Hawkins spoke to me about one such reflex effect on himself, having to do with his reactions to the first atomic bomb test on July 16, 1945. The Trinity test was conducted about two hundred miles south of Los Alamos, in an area intersecting the northwest corner of the Alamogordo Bombing and Gunnery Range. This wilderness had been known for centuries as La Jornada del Muerto—Dead Man's Journey. Hawkins explained to me that although maps often identify a large desert area east of the Rio Grande as La Jornada, it was originally a particular small stretch of land. The area was so named because when some of the early Spanish explorers first came up the Rio Grande, they discovered a big bend in the river. They calculated that, by leaving the river, they could cut across for a good many miles, thus avoiding a longer route through what they believed was the territory of hostile Indians. The Spaniards set out across the region without realizing it contained no water, and many died of thirst.

Robert Oppenheimer christened the test and its desert site "Trinity," and I had read about the symbolism of his choice. In a 1962 response to a query from General Groves regarding the code name's origins, Oppenheimer wrote that he suggested the name, but it was not clear to him exactly why. He told Groves of a poem of John Donne's that he loved, written just before the poet's death:

> . . . As West and East
> In all flatt Maps—and I am one—are one,
> So death doth touch the Resurrection.

Oppenheimer went on to remark, "That still does not make Trinity; but in another, better known devotional poem Donne opens, 'Batter my heart, three person'd God.' Beyond this, I have no clues whatever."[4] Hawkins told me that he distinctly recalled Oppenheimer discussing this verse with him at the time of the bomb test.

The philosopher also said that about six weeks before Trinity he had been "perfectly frightened" about the possibility that the fission reaction might ignite a thermonuclear reaction in the atmosphere. Although the scientists had assured themselves that this would not occur, Hawkins remarked that he was "just naive enough to not have been through this whole persuasive calculation that the physicists had done." And he was distressed by what he described as the casual way in which the possibility was dismissed.

As early as 1942, at a Berkeley meeting of bomb theorists, Edward Teller had raised the problem that an atomic weapon might ignite a thermonuclear reaction in the atmosphere. At that time Oppenheimer had been concerned enough to place an urgent call to Compton, director of the Chicago Metallurgical Laboratory's efforts to produce the chain reaction. Compton recalled Oppenheimer arriving at the door of his Michigan vacation cottage the very next day.

> What his [Oppenheimer's] team had found was the possibility of nuclear fusion—the principle of the hydrogen bomb. . . . To set off such a reaction would require a very high temperature. But might not the enormously high temperature of an atomic bomb be just what was needed to explode hydrogen? . . . [W]hat about the hydrogen in sea water? Might the explosion of the atomic bomb set off an explosion of the ocean itself?
>
> Nor was this all. The nitrogen in the air is also unstable, though in less degree. Might it not be set off by an atomic explosion in the atmosphere?[5]

Hans Bethe later explained, "In fact, a few days after Teller first proposed his idea in 1942, I proved that it was impossible. As part of their work at Los Alamos, Konopinski and Teller worked out an even more thorough proof."[6]

In summer 1945 David Hawkins knew of the calculations showing the impossibility of such a thermonuclear reaction. But because he was beginning work on the Los Alamos project history, he decided to look more deeply into the question. Hawkins told me,

> I went to everybody to talk about it who was thoughtful about the virtues and the limitations of physical theory. I talked to Phil Morrison about it, and Phil said, "Well, if there was some nonlinear effect that we

don't know about—possible. How can you rule it out?" I talked to
Robert Oppenheimer about it who gave me a very straight response
and said, "Well, we know a great deal about the repulsion forces of the
nuclei of atoms and to get two nitrogen atoms to a point where they
would interact would take an enormous amount and concentration of
energy." I asked Edward [Teller] about it, and he explained to me the
calculations which he had done—he was the author of these calculations—and it had been a matter of serious concern briefly, but then
Edward's calculations seemed to show that it was way off scale and it
couldn't happen. But the question remained whether the calculations
were adequate.

And, being a philosopher, I couldn't help thinking of a parallel which
is known as Pascal's Wager—Pascal being a devout Christian of a very
passionate kind nevertheless was also a scientist and mathematician.
And he said, "Assign to the probability that the Christian story is true
a probability which is very small. Even so, the gain from accepting it is
almost infinite and the loss from accepting it, if it is false, is very small.
Therefore, you should accept it." And this he called "the Great Wager."
This was the acceptance of an article of faith. I said, "This is like the
Great Wager in reverse."

Hawkins went on to tell me that Teller said, "Oh well, David, worse
things could happen," and that it would be so fast nobody would ever
know it. Remembering that Bethe's physical knowledge, intuition, and
insight were probably better than anybody else's, Hawkins told me, "I
think I did talk to him about Teller's calculations, but I think at that
time he just rather dismissed me without any full explanation." Not
cheered by what he was hearing, Hawkins kept asking questions.

I talked to [Richard C.] Tolman, the great man, one of Groves's big
advisers who was visiting—I think it was Tolman, it might have been
[James B.] Conant—they both came and I didn't make any notes about
it.[7] There I got the answer, "Well, it's too late to reconsider now."

Hawkins laughed as he continued,

So I was not happy. But I've not really told this story much because I'm a
little embarrassed. I seem naive compared to the people who knew

more than I did. But you see, I did have the Great Wager question to throw them and they had no answer to it, of course. And that was for me, this is a personal note, that fright made me have a perception of the enthusiasm for the upcoming test that I felt somewhat alienated from. And that's why I decided not to go to the test, personally. And I made the plausible argument, somebody who has been more involved in this than I have can go in my place, and I guess somebody did.

At about 5 o'clock on the morning of the Trinity test, Hawkins went to the technical area of the laboratory where a Women's Army Corps (Wac) teletype operator was waiting for a message from the Alamogordo desert. He asked her if she had any news, and she replied she had heard nothing. Hawkins was gazing out the south window of her office when he saw the sky become suddenly brighter. Turning to the Wac, he joked that she would be hearing in a few minutes. He told me that he was probably the only person at Los Alamos who saw the light from Trinity.

The light went up, but it went off. It was a dawn sky, it was not daylight sky. I just saw the whole southern sky light up and then that was it. It was the immediate flash of light. I didn't see the fireball or anything like that. It was the flash of light that preceded the expansion of the fireball.

Hawkins confided that he felt even more alienated as the scientists came back from the test filled with excitement about their success. Even those who had expressed serious concerns about the use of the bomb and its wider significance were thrilled that they had built the atomic device and it had actually worked. He remembered being unable to share their technical enthusiasm.

As I listened to Hawkins, I thought it made perfect sense that the scientists would be completely focused on the technical work, divorced, in a sense, from its larger implications. Scientific research, after all, involves verifying theory in experiment, and the Trinity experiment had been a more daunting challenge and a greater success than anyone could have imagined. For the scientists, it was a professional achievement of tremendous proportions. As Philip Morrison noted fifty years

after Trinity, "For most of the 2,000 technical people at Los Alamos—
civilians, military and student-soldiers—that test was the climax of our
actions. The terrifying deployment less than a month later appeared as
anticlimax, out of our hands, far away."[8]

I was amazed when Hawkins went on to tell me that the atomic test
had been described ahead of time in a paper called "Phenomenology:
What It Will Look Like," produced by one of the Trinity project groups.

> The expanding fireball, the blue glow, everything carefully documented
> from theoretical calculations, and the excitement was partly that the
> prediction was so accurate. It was as though you knew ahead of time
> what was going to happen, but when it actually did happen it was
> beyond your ability to accept it—very complicated. The term "phenom-
> enology" was both apt and inept because the real phenomenology was
> in the reaction that was experienced. But the description was accu-
> rate—the description of the facts, not of the human reaction. And of
> course it's that latter one which is really important.
>
> I think the excitement when they came back was sort of the first
> reaction to that. And I'm sure many people thereafter said, "Oh my
> God," and some people said it at the time. As though we foolish
> humans hadn't really known in advance what we were doing, that kind
> of reaction. And that's what tragedies are made of. Anyway I think my
> own dissociation from all of this had been developing.

At the moment that the scientists transformed their theories into the
reality of the bomb, it was clear that their science ceased to exist solely
on the level of ideas. Their achievement was not isolated in the laboratory but had occurred amid the manifold and complex historical inter-
sections of extraordinary scientific discovery, a long and terrible war,
national politics, and fundamental shifts in global relations. The scientists were sure that the bomb would not set off a thermonuclear reaction that would ignite the atmosphere. They were able to describe,
before the fact and with great accuracy, the phenomena of the Trinity
test. However, they were unable to predict the shock waves of their creation in the social world.

After a full morning of conversation, Hawkins and I took a break for
lunch at a neighborhood Chinese restaurant. There I learned more

about the personal phenomenology of his experience of Trinity. He told me that from about age ten he had grown up in the village of La Luz (Nuestra Señora de la Luz, Our Lady of the Light), just a few miles north of the town of Alamogordo. Hawkins's father had been a successful pioneer lawyer. In 1885 he had left Tennessee for southern New Mexico and nearby El Paso, Texas. When he retired, the family moved to La Luz. There he became a gentleman farmer, though with limited success. Hawkins recalled that one of his father's interests was science. He became a member of the American Association for the Advancement of Science and was fascinated by the energy of the atom, then known only through the discovery of radioactivity.

Hawkins also told me that when a Los Alamos group was scouting for a site for the test of the implosion bomb, he led them to a desert area he and his friend Berlyn Brixner had explored in their teens. Hawkins recommended Brixner for the job as a photographer at the Los Alamos lab. Unbeknownst to Hawkins until fifty years later, his childhood friend had also scouted the site for the bomb test. Some time after my interview with Hawkins, I met Brixner, who, like my mother, was a member of the wartime lab's optics group. He made many important innovations needed to capture the various stages of the scientific and military work on film. It was Brixner who had taken the series of Trinity photos I studied as a child. And he remembered my mother.

The Trinity site, like Alamogordo and La Luz, is in the Tularosa Basin, an ancient rift valley, but some fifty or sixty miles to the northwest. When I asked Hawkins if his personal connection to the site had influenced the final decision to adopt it, he said, "Only by introducing them to it." The area was big and barren enough. The Alamogordo Bombing and Gunnery Range was already in possession, and General Groves could simply commandeer a section for the Trinity test. Hawkins said that after that first scouting trip he returned to the area only once, a year after the war. Then he saw the bomb-print depression and the melted sand that remained.

While at the time Hawkins made it known at the laboratory that someone more vital to the test should go in his place, he admitted to me that his attachment to the desert landscape of his youth could have been part of what kept him from witnessing the historic event. Perhaps he did not want to see his "patria chica" (little homeland) put in the service of such purposes. And, in an imagined conversation, he told his dead father, "I know this is not what you would have wanted the power of the atom to be used for."

On returning from lunch, I decided to follow up on some of my questions from the morning's conversation. I wanted to know Hawkins's thoughts about the social responsibility of scientists for the applications of their creations. He answered by referring to the atomic scientists' postwar political movement. However, he asserted that to whatever extent atomic scientists may have bad consciences, this did not explain their subsequent political commitment. He framed his response in terms of Oppenheimer's statement about the scientists having "known sin."

> He wasn't talking about them saying "Mea culpa, mea culpa," but he was talking about them acknowledging responsibility and the commitment that came with the responsibility: I'm in this thing, I can't get out of it anymore, it's mine and it's mine politically, not just as a techno-scientist.
>
> At the time it was very unusual for anybody to say that scientists, as scientists, should or could have an independent political voice. I wrote a piece in the New York Times when I was still at Los Alamos called "Should the Scientist Take Part in Politics?"[9] And the argument was yes, but there was historical reason, which is that the time lapse between scientific discovery and technological applications was originally so long that the scientists could not have any responsibility. I think it was Faraday who was asked about his discoveries of electromagnetism. Somebody said, "What use is this?" and he said, "What use is a newborn baby?"

Hawkins explained that historically, in the cases of many nineteenth-century scientific discoveries, technical and industrial applications were not developed until long after. However, over time, the delay between discovery and application shortens. Wireless radio communications came two or three decades after discoveries that took place at the end of the nineteenth century. During World War II, the development of radar and nuclear weapons took only a few years. Enrico Fermi and his colleagues achieved the first controlled nuclear chain reaction on December 2, 1942, four years after the discovery of fission. Two and a half years later, the Los Alamos scientists detonated the first atomic bomb, and within three more weeks, the weapon was used on Hiroshima.

> So the very development of science, on one hand, and technology, on the other, makes them into Siamese twins, if they weren't originally. In fact, all of the noblest traditions of science are independent of this and most scientists don't like to consider themselves as inventors. It's this combination of research which is directed by narrow economic goals that becomes the real technoscience. So the argument is as the gap gets shorter, the inevitable commitment to the consequences becomes more meaningful. But [at the end of World War II] it was considered a very debatable point.

This debatable point can be seen at the heart of Niels Bohr's attempts, before the end of World War II, to influence postwar nuclear policy. During the course of my two-day interview with Hawkins, our conversation often looped back to this question. While we did not follow a linear progression on the subject, Hawkins's insight into the importance of the relationship between Bohr and Oppenheimer, and its consequent influence on other Los Alamos scientists, was an underlying, recurring theme.

Oppenheimer considered Bohr's vision that atomic energy should be used for the benefit of humanity and not as a menace to civilization more important than any technical aid he might bring to the bomb-building project. I knew of Bohr's question to Oppenheimer about the bomb, "Is it big enough?" He was asking if nuclear weapons, because of their unprecedented destructive power, would mean the end of the institution of warfare. As I listened to Hawkins, this question no longer referred to theoretical concepts about the meaning of the bomb but was grounded in the relationship between two living, breathing men in an extraordinary historical moment.

Hawkins told me about Bohr's visit to Los Alamos after meeting with Churchill and Roosevelt. The Dane had undertaken his independent mission to meet with the Allied leaders in order to convey the urgent need for international control of atomic energy. He did so as a world-class scientist with colleagues on all sides of the European conflict. He was convinced that a crucial element in avoiding a postwar nuclear arms race was to inform Stalin of the Manhattan Project before war's end. Hawkins believed, as did others, that although Bohr's meeting with Churchill had been a disaster, Roosevelt had "seen the light." Hawkins conveyed to me the sense of hope that Bohr's 1944 visit brought to many in Los Alamos and to Oppenheimer in particular. Bohr's question to Oppenheimer, "Is it big enough?" was Oppenheimer's "basic deep pitch from his tutor, his boss, from his mentor." This philosophy was embodied in the concept of nuclear pacifism, a phrase invented by Philip Morrison, which Hawkins remembered using early on. Apparently, Hawkins reflected, Roosevelt always strongly opposed this concept. "He shared Churchill's scorn of this crazy physicist," but the scientists did not realize it at the time.

Hawkins has written that for Oppenheimer, the meaning of nuclear weapons was "vastly larger than the weaponeering frame that confined it—and him. . . . Ironically, weaponeers could light the path brightly; danger was also opportunity. The very magnitude of nuclear destructiveness would force the nations to control nuclear weapons." And Oppenheimer himself reflected twenty years after the fact, "Bohr at Los Alamos was marvellous. He took a lively technical interest. . . . But his real function was, as least for many of us, a very different one; he made the enterprise, which often looked so macabre, seem hopeful."[10]

Hawkins recalled that to some in Los Alamos Bohr seemed like the messenger god Hermes. He explained,

> When Bohr came to Los Alamos we were all full of hope, of optimism. [We thought,] if the great Roosevelt really understands this, then that's a wind blowing in the right direction. And we were not disillusioned during the war about that. Roosevelt died. Truman took over. I remember Oppenheimer as saying, "Well, Roosevelt was a great architect, perhaps Truman will be a good carpenter," which was not a bad remark, but on this subject Truman had no greater insight than anybody else.

Hawkins thought that for Truman the atomic bomb was "the winning weapon."

Another incident that took place during one of Bohr's visits to the Los Alamos lab sheds light on Oppenheimer's personal attachment to his teacher. Hawkins remembered,

> Robert Oppenheimer was taking Niels Bohr back to Fuller Lodge, where he was staying when he arrived, walking past what we called "the pond" [Ashley Pond], a little pond of water about one thousand square feet, not much bigger. And Oppenheimer had him by the arm walking him back. I was going the other way. And then coming back, Robert was coming past me again, at almost the same place and we stopped and talked. It was winter, and he said to me, "While I was walking here past the ice it occurred to me that if he [Bohr] slipped and injured himself it would be a terrible tragedy. I was so afraid." This was his attitude toward his mentor.
>
> And the next day, I think it was, he [Oppenheimer] beckoned to me and took me into his office and pulled out the file drawer and showed me—made me put on gloves—and showed me a piece of paper, two pages. And it was a copy of the paper Bohr had given Roosevelt. And it was virtually the same thing that he later printed. I was shown this precious document—and obviously Bohr had left this with Robert, although I never heard him say so. And the implication was that Roosevelt had fully understood. And this was a great source of joy and optimism. But anyway it's another example of his attitude toward Bohr and of Bohr's basic belief reflected in that strange question "Is it big enough?"
>
> It's interesting. We all lived under this illusion, you see, for the rest of the time at Los Alamos, that Roosevelt had understood. And we knew after Truman that he probably didn't understand. We thought there was some basic U.S. policy commitment already. And at the end of the war it became super clear that that wasn't the case.

In his memorandum to Roosevelt, Bohr had written, "Without impeding the importance of the project for immediate military objectives, an initiative, aiming at forestalling a fateful competition about the formidable weapon, should serve to uproot any cause of distrust between the powers on whose harmonious collaboration the fate of coming generations will depend."[11] Some have characterized Bohr as a hopelessly idealistic scientist, meddling in world affairs. Others consider him a visionary, possessing a profound understanding of the impending dangers if world powers did not begin to relate to each other in fundamentally different ways.

Oppenheimer asserted that Bohr's failed attempts with Roosevelt and Churchill show "how very wise men dealing with very great men can be very wrong." McGeorge Bundy wrote that the meeting between Churchill and Bohr "is one of the first and most famous failures in the arduous process by which scientist and statesman seek to understand each other on nuclear dangers." Martin Sherwin rejected interpretations of the encounters as historical tragedy—the notion that the statesmen would have been sympathetic had they only understood the scientist's message. According to Sherwin, Roosevelt and Churchill understood the physicist all too well, and fundamentally opposed his ideas, which ran counter to their dedication to a postwar Anglo-American monopoly on atomic power. Bohr's biographer, Abraham Pais, summarized, "Was Bohr's attempt naive? Did Churchill's and Roosevelt's lack of response show lack of wisdom and foresight? My own answer to these questions would be no. Times were simply not yet ripe."[12]

Through my discussions with Hawkins, I gained a deeper understanding of the degree to which Oppenheimer's and many Los Alamos scientists' hopes for international control had rested on the great man. Yet Bohr had seriously misjudged his meeting with Roosevelt, and he, Oppenheimer, and other project scientists were completely unaware of his utter failure. I wondered whether the hope for postwar control was a way for the weapons builders to make sense of their actions. I could understand the logic of arguing that the bomb had to be used in all its destructive power as a weapon in World War II or it would be used in World War III, fated to be an all-out nuclear war. But something remained deeply disturbing to me about this calculus.

I told Hawkins that two thoughts came to my mind. The first was that conceiving of the weapon in this way meant that the eventual victims of the bombs, whomever they might be, were in essence con-

demned to be sacrificed on the altar of the scientists' perceived cause—
the deterrence of World War III. Yet, on reflection, I realize that history
may prove that this was indeed the case.

I conceded that my second observation could be made of many historical figures. There was something frightening to me about a man like
Oppenheimer, whose convictions about the ultimate meaning of the
bomb were so tied to his own vision of himself as a heroic figure destined to bring this knowledge into the world and to reveal to humanity
these great changes. His relationship with Bohr just added another
dimension—Oppenheimer would be the one to fulfill his scientific and
philosophical father's vision.

On several occasions Hawkins suggested that I cultivate a broader
view of Oppenheimer. He told me, "Oppenheimer's whole career, when
you look at it, is a kind of Greek tragedy—a tragedy of power and of the
overestimation of power by the person who has it." And, Hawkins
added, when Oppenheimer went to Washington in the early days after
the war, he was unwilling or unable to accept the support of his junior
allies. "And that's always courting danger." To illustrate his point,
Hawkins told me of seeing Oppenheimer, Philip Morrison, and fellow
physicist Robert Wilson in 1946, after the failure of the United Nations
to establish international control of atomic energy.

> [Oppenheimer] had to go it alone in Washington. He welcomed the
> Federation of American Scientists as a positive move in the right direc-
> tion, and he gave us a lot of moral support—so did other senior people
> who were reluctant to join, it wasn't their métier to be members of an
> organization, a political organization. And we were all a little distressed
> by this. He didn't even tell us what he was doing in Washington.
>
> The three of us [Hawkins, Morrison, and Wilson] were at a meeting
> at Princeton University's Tercentenary, and the first [postwar] interna-
> tional physics conference was going on, to which there were invited a
> Russian physicist or two. So this was quite an occasion. And I was there
> not being a physicist, I was an observer, a friend of various participants.
> And this memory is very vivid in my mind: we were in a small room.
> Bob Wilson, Phil Morrison, and I were talking to Robert Oppenheimer.
> And he said, "You know, now that the proposal for U.N. international

control has failed, you should just really abolish the organization because you have a perfectly clean record and if you go on you will inevitably make mistakes and your record will be tarnished"—very strange argument.

And we all rose up against this, and as we started to argue, someone came to the door and said, "Dr. Oppenheimer, you have a phone call from Washington," so he disappeared, we regrouped our forces, so to speak. The three of us who all had been his subordinates and admirers said, "He can't be right, he's crazy, that's wrong." And he came back and we gave him a fresh argument against it. Then he was called out again and the same thing happened. Here were these three junior devotees recovering in their own mind from something like worship. Finally he left, having realized that we had not accepted his argument. We looked at each other and Bob said, "Gee, we really stood up against him, didn't we?" I think that's something for your notebook.

On first hearing this, I assumed that Oppenheimer had encouraged these younger disciples of Bohr to essentially abandon their vision. But Hawkins later told me that this was not the case. Rather, Oppenheimer had meant that the well-fought battle for international control of atomic energy had been lost. The Federation of American Scientists had failed to achieve its purpose, and the members had been politically virtuous and should therefore disband. Hawkins also made a further observation about his old friend.

You know, when you get way away from this, here's a man who had resonated so with the bloody and fearsome Hindu scripture before any of this came about. And who then, somehow, saw himself in this image. And then met the disaster that he finally met partly because of an understandable and incurable devotion to a cause, not unmixed with personal arrogance. It really is a tragic story.

I understood that Hawkins was referring to Oppenheimer's identification with the Bhagavad Gita. After witnessing the Trinity explosion, Oppenheimer spoke the verse "Now I am become Death, the destroyer of worlds."

✷ Pacific Memories II

September 8, 1996
Denver International Airport restaurant after Hawkins interview

Eating breakfast and waiting for plane

I take the shuttle bus from Boulder to the airport. It stops at several hotels and motels and the seats fill up. A man sits next to me; his wife sits a few rows in front of us. I surmise he is probably in his seventies and wonder if he is a World War II veteran. He asks if I've been visiting family in Boulder, and I tell him that I have been working. We begin making small talk. He tells me that he and his wife went to several historical museums in Boulder and he thought they were quite good. He adds that he considers himself something of an amateur historian, having been a history major.

Mr. Porter [a pseudonym] says that he studied European history at college because he had a great professor who actually urged him to stay on to get his doctorate. I hear regret in his voice as he tells me that he does not know what would have happened if he'd gone that route. But as it was, he became a teacher and then an elementary school principal. He and many others at college after the war were on the GI Bill and did not want to stay in school but were anxious to get back to work in the world.

I ask him where he was during the war, and he says that he was a Marine stationed in the Pacific. He makes a specific point of telling me that he was not in combat but received training to become a specialist in decoding Japanese messages sent between ships, later between ships and bases. I tell him I have read some books on the codes, and at about this point in the conversation I say I am doing oral historical research about the Manhattan Project. I ask where he was at the end of the war. He replies that he was back in Honolulu getting training—that he was actually playing baseball when the announcement of the surrender came. He says they started drinking beer and finished the ball game drunk.

Mr. Porter says he and his buddies were very scared of the possibility of an

*invasion. Even after Hiroshima and Nagasaki they had no idea whether the
Japanese would give up and did not let down until they heard news of the
actual surrender. Then he was stationed on Kyushu for the occupation where
he was charged with watching for guerrilla activity, but there was none. Like
Mr. Hamilton, the veteran on the plane, he makes a distinction between the
Japanese in power and the ordinary Japanese people. And he talks about the
rise of Japanese militarism in the 1930s.*

*He tells me the invaders are always at a disadvantage—the Japanese
would have just stayed in their holes and shot at them. Then he says, with
some hesitation, that he thinks without a doubt that the bomb saved countless American lives. (He is tentative—is he wondering what I will think? I got
the same sense from Mr. Hamilton.) We discuss the invasion, the bomb, the
controversies during the fiftieth anniversary. He says he has four children,
two conservative, two liberal, and they argued during the fiftieth anniversary
about the decision to use the bomb. I tell him I don't think this issue necessarily breaks along party lines—that my liberal uncle was stationed in the Pacific
during the war and has no doubts that the use of the bomb was correct.
Porter says he knows it killed a lot of civilians, but we would have lost
so many men in an invasion. We discuss the various sides of the argument,
including that the fire bombing of the Japanese cities did not bring surrender.
He says that this caused him and his buddies to remain afraid of invasion,
even after the atomic bombs.*

*I tell him about Dad supporting a demonstration of the bomb. And I say I
am unsure whether a demonstration was realistic but do think a real chance
was lost by not offering to allow the emperor to stay. He counters that uncon-
ditional surrender prevented that, and I reply that without the atomic bomb
first, perhaps the American people would never have accepted the emperor's
retention. He says something to the effect that we could not let them off that
easy. I remark about the need for revenge, not as an assertion, but as a point
of discussion. It occurs to me that after the atomic bombings revenge was satisfied. I say that I think the only real possibility of avoiding the bombings was
our promise to retain the emperor and a stronger warning about the nature
of the bomb.*

*He shows me his only war wound, a small scar on a finger of his left hand.
He broke a bone playing baseball, and it pierced through the skin. Until it*

healed he could not do his decoding work on the typewriter. I say, "You were
lucky," meaning lucky not to have been seriously wounded. He agrees he was
lucky but means something different: the injury occurred around the time of
the battle of Peleliu and he could not go. The fighting on the island was so
fierce that some who went to do decoding work had to go into battle. It was
so bad, he tells me, "everyone there received the Purple Heart."*

My encounters with two veterans of the Pacific war form bookends to my interview with David Hawkins. Mr. Hamilton and Mr. Porter were very different from the vengeful, vehement, angry veterans portrayed in the press during the Smithsonian *Enola Gay* controversy. In contrast, they were both reticent to express an opinion that might differ from mine. Recalling their participation in the occupation of Japan, they expressed sympathy for the Japanese people and empathy for their plight.

I got a strong sense of Hamilton's attachment to his friend, Lewis, killed in the battle of Santa Cruz. With Porter, what struck me was the bond within a group of men—his buddies. When not speaking of situations specific to him, he always said "we"—"*we* were afraid of the invasion, *we* went to Kyushu." Neither of the men insisted on the rightness of their views; both were aware of the ambiguities. As I listened to Mr. Hamilton and Mr. Porter, they seemed vulnerable, even delicate. Perhaps it was the anonymity of our meetings that allowed them to share a chapter of their life stories with me. Perhaps it was the passage of so many years.

These are the men of my parents' generation who went to war and came home to raise families and make careers. I grew up with many of them but never heard their stories. I remember being deeply moved in

---

* Peleliu is a seven-square-mile coral island five hundred miles east of the Philippines. In September 1944, 16,000 U.S. Marines landed, expecting to take the island in two or three days. The plan was to protect General MacArthur's flank as he advanced on the Philippines. More than 10,000 Japanese were dug in, and sixty-eight days later the Americans prevailed. The battle was one of the bloodiest of the Pacific war. Reports I have read place the American dead at between 1,200 and 1,500. Many thousands more were seriously wounded. Most of the Japanese died. The island's capture did not contribute significantly to MacArthur's victory.

1994, during the fiftieth anniversary of D day, when I learned that so many veterans had remained silent about the horrors of the invasions for a half century. Perhaps the powerful symbolism of the victory's golden anniversary, combined with its proximity to the end of the century, brought it all full circle.

CHAPTER 5

Robert R. Wilson,
the Psyche of a Physicist

In October 1996 I traveled to Ithaca to conduct several interviews at Cornell University. I arrived on a cool, damp day and spent the afternoon and evening alone at the hotel completing preparations for the week's interviews. As I worked, I looked out on the steeples of Ithaca and the clock towers of Cornell.

The next afternoon the university's venerated Nobel laureate, Hans Bethe, gave a colloquium on the making of the atomic bomb. I had arranged to meet physicist Kurt Gottfried, a former colleague of my father's, before the lecture. As I walked down the third floor hall of Cornell's Newman Laboratory checking for Gottfried's office number, I looked through an open door to see Bethe's slender shoulders curved over his desk. I hesitated for a moment, wondering if I should greet him, but did not want to break his deep concentration.

I had never met Gottfried, but he and my father had become good friends in the early 1970s, when they both worked at the Conseil Européen pour la Recherche Nucleaire, or CERN, a laboratory in Geneva where several nations collaborate on basic research. I was looking forward to meeting another scientist who had known my father in his prime. We visited for about half an hour and then headed to the lecture hall, where a large crowd was gathering outside. When we entered the room, Gottfried spotted Robert Wilson and introduced us. I immediately recognized the Manhattan Project experimental physicist from photographs and film, but now, instead of brown, his hair was a shock of white. Looking into Wilson's handsome, square face, I thought I discerned traces of the delicate fault lines of serious illness.

As it turned out, Wilson was directly responsible for Bethe's upcoming lecture. In 1995 he had received an award administered by the American Physical Society given specifically for outstanding work linking physics to the arts and humanities. Wilson, both a master builder of accelerators and a sculptor, was an ideal recipient. Part of the award was a grant to the physics department of his choice for a lecture series of interest to the public as well as to the local scientific community.

Gottfried was involved in the presentation, so we separated. I fought the impulse to hide in the back and took a seat in the front of the hall. The chairs and then the aisles quickly filled. The audience was a mix of professorial-looking men in their sixties, seventies, and eighties and young students, men and women. It was clear that Hans Bethe is deeply respected at Cornell. He is credited with making the university's physics department world class, and many physicists acknowledge him as instrumental in their personal success. The affection and pride that they feel for the old gentleman is palpable. I watched as he shuffled confidently to the podium, surveyed the audience, and, with the slightest smile, began his lecture.

During his presentation, Bethe conveyed the enormous dedication and precision with which the entire Manhattan Project experiment had been conceived, calculated, constructed, and executed. This refinement of human thought and ingenuity was in sharp contrast to the gross outcome—bombs that killed and maimed so cruelly and indiscriminately.

Early the next morning I completed my preparations for the interview with Robert Wilson. During my research, I had learned of his personal views on the decision to use the bomb, his admiration of Niels Bohr, and his role in the creation of the Federation of American Scientists. A native of Wyoming, Wilson had been inspired in his study of science by a high school teacher. He was one of Oppenheimer's first recruits to Los Alamos and had worked closely with the director on the facility's early organization. As head of the Cyclotron Group, he was the youngest group leader at Los Alamos. After the war he taught at Harvard and, in 1947, came to Cornell to direct the Laboratory of

Nuclear Studies. In 1967 he became the first director of Fermilab, the national accelerator center in Batavia, Illinois. Wilson has written about his wartime experience, and there were several topics I was hoping to discuss with him. One was his questioning of the continued development of the bomb after the Germans were defeated, the second was his experience of an epiphany at the Trinity bomb test, and the third was what he described as a sense of betrayal on learning that the atomic bomb had been used on Hiroshima.

Wilson came by the hotel at midmorning, and we drove back to his home for the interview. As we entered the front door, his diminutive wife, Jane, was slowly descending the staircase to greet me. I was pleased to meet her, as I knew of her work as an editor and writer. I had been searching for a copy of her 1974 volume, *All in Our Time*, reminiscences of twelve Los Alamos scientists. Later in the interview, when I complimented Wilson on one of his articles, he responded with a smile, "Jane is responsible for a lot of my writing."

The Wilson home was filled with light and with the graceful sculptures, large and small, that Robert Wilson had created over the years. And I noted the understated beauty of the furniture and art objects. Outside, several porches overlooked the glacier-formed cliffs and waterfalls that surround the hundred-year-old house. The three of us sat in a little enclosed sunroom that was perched like an aerie above the magnificent rock formations. A few weeks earlier the middle-of-October leaves had been ablaze in magentas, scarlets, oranges, and yellows. Now they were fading to brown and falling to the ground. But I was not disappointed to look out at the last of the season's finery. As a child, I had reveled in the glories of the eastern autumn. Now transplanted to California, I was happy to witness it again in any form. I discovered that I could look deeply into the dying colors and perceive the beauty of times past.

We visited informally for a few minutes. Jane Wilson recalled my father being at the house for cocktails when he had come up from Brookhaven to give a talk. They would not have invited him had they not met previously, she said. So we sorted through times and places in an attempt to discover whether they might have also known my mother. As

Figure 13. Robert and Jane Wilson, Berkeley Radiation Laboratory, 1940.

we chatted, I thought I heard the signs of speech or memory problems in the slight hesitation of Robert Wilson's voice.

We settled into the interview, and I asked Wilson what he had thought of Hans Bethe's lecture the previous afternoon. He told me that, as usual, Bethe had done a superb job. He remarked that he had actually forgotten some of the Los Alamos history and enjoyed having his memory jogged by the talk. And he wondered aloud whether Bethe had referred to some details he had never actually known—"Just things he brought up that I wasn't sure I knew about. Although I can't remember them, it was so long ago."

Wilson reached farther back in time to put his views of the lecture in the context of his personal history. From this vantage point, he had a different perspective on the many project successes that Bethe had enumerated. Wilson explained that his forebears had been Quakers and that he was a pacifist when the war began.

> So it was quite a change for me to find in fact that I would be working on this horrible project. On the other hand, I certainly understood what that entailed and it would probably be used—if we were successful. I found one thing interesting: Hans always seemed to say, "Well, we were *successful* in doing this and *successful* in doing that." I always hoped we

would not be successful. I think—he has a way of talking about working on a scientific project—if it comes out the way you expected it to come out, well, that means that you are successful. And I kept hoping that we would find something [that would show an atomic bomb was not possible]. And we could have. For example, he talked about the delay of neutrons and if all the neutrons had been delayed there would have been no bomb. And so, I thought *that* would be a successful ending.

Thus Wilson revealed the first of his many reflections on the very meaning of "success" and "failure" when considering the troubling legacy of atomic weapons. If the scientists had been proven wrong and the bomb had been impossible to make, he would have seen it as a success. I wondered silently if that meant the scientists' success was a human failure. I asked Wilson if, while working on the project, he had consciously hoped that something would go wrong and they would all be out of it.

I guess. It's complicated. Certainly in the beginning we were losing the war to the Nazis and that seemed just a horrible thing and if we could win the war somehow—the alternative was we could have had a thousand years of Nazism or three hundred years or ten, but any length would be too long.

Jane Wilson emphasized that the social atmosphere of the war years had profoundly influenced their choices. She explained that they were afraid not just of the events in Europe but of those occurring in this country. She spoke of prominent figures supportive of Hitler's goals, including Charles Lindbergh and Henry Ford.

The success of Hitler was breeding all sorts of very vocal groups in the United States. There was no hope that we would be an isolated little island of democracy in a fascist world. Of course, nothing breeds success like success. So if Hitler had finished his conquest of Europe we would have been in grave danger, grave danger. Not from conquest, but from within.

Although aware of the fascist and racist forces in this country, I expressed some interest in hearing that the actual fear of conquest from within had been so strong. Jane Wilson pointedly asked, "Is that a new thought to you?" When I replied that it was and admitted that this showed my generational ignorance, she quipped, "It sure does," and went on to explain that today it is difficult to imagine what the atmosphere in the country had been in 1943.

> It really looked as if the forces of fascism on both fronts were just winning. MacArthur had departed from [the Philippines] saying, "I will return." But he had departed, and you just had that sense that there was no hope. [It was] very different—not at all like our cold war with "The Russians are coming." My foot! I certainly didn't believe it. But this was, I think, much realer.

Robert Wilson then augmented her point by saying that at the time there was endemic anti-Semitism in this country. I did know about that. My father attended Northwestern University on a scholarship. On the first day of classes, the director of the engineering department called him into his office and said that while they were pleased to have such a bright young man as a student, my father had to understand that he would never actually work as an electrical engineer. The profession was closed to Jews. And it was true. After graduation my father worked for the telephone company repairing lines. Wilson now told me of anti-Semitism he had seen at Princeton, but the war had changed everything.

> That stopped it. But had the war ended the other way, all of those anti-Semites would have just gotten stronger in the way they were. So I think whether we won that war or we didn't win it did make a difference. And once I had concluded that, then it was extremely important to win the war.
>   Now then, they [the Germans] were working on the bomb too, and so that doubled the reason to work on this. Because if they had gotten the bomb, and they might very well have before we did, then we would have lost the war. I thought he [Bethe] was right to say we were successful in doing something. I guess that I would have, this is from a personal point of view, as a pacifist manqué, I would have preferred not to be so involved with what we were doing. And after a while of course it became clear that we were not losing the war unless the Germans came

up with something that they had well hidden, and that might have been true too. Just as we surprised them, they could have surprised us. So it's complicated.

But at the end of the war, I felt that I served my time and that I wasn't going to do that *anymore*. Because I couldn't complain about the bomb being used the way it was, I'd been working toward that end and couldn't of course have any effect on how it was used. I was one inconsequential person. So I would rather that we hadn't. I'd rather feel that God's in his heaven and all's well with the world. There are a lot of things that don't come out the way I'd like them. But anyway, I felt that I had done my bit, and then I felt that we had essentially been promised throughout the war that there would be a United Nations. When the cold war started, then we seemed to be starting all over again with a new enemy. As though we *had* to have an enemy. Well, I felt that I wasn't going to work *anymore, ever,* on making bombs.

When I commented that I knew others who had been pacifists at the outset of the war, Wilson added, "I wasn't a very good kind of pacifist. I suppose that a really good pacifist wouldn't have worked on it even so. They would continue to be pacifists [even] when Nazism came. That's the way they're supposed to be." But he emphasized that this was not the point he was trying to make.

What I wanted, really wanted, to say was that many people continued to work on the bomb [after the war]. Hans did. And I didn't. I took the stand "That was it." I wasn't going to do it anymore. But how it turned out was that Hans continued to be an expert. When he spoke out, just like last night, he would say what was the truth and what had to be done and what was possible to do. Whereas since I was no longer connected with it, I was an amateur—almost lost any expertise that I had previously. I lost it all, so I had no effect on what was happening.

It's an interesting way it turned out, that the people who could have something to do with war or peace continued to be effective. Hans has continued to be a strong voice for peace in the world where I was a no voice.

Wilson's self-portrait was in contrast to my own conception of him as someone who had taken strong action based on his inner convictions during and after the war. I told him that I did not know if I could agree with his characterization. I thought I saw Jane Wilson begin to speak, but when I asked what she was going to say, she simply replied, "I don't know." Yet Wilson was making an important point—taking his stand had meant that he would have no say in the matters about which he cared so deeply. As the late I. I. Rabi commented, "You had to be inside the government if you wanted to have influence. . . . Since there was all that secrecy, you couldn't know what you were talking about unless you were part of it."[1] Wilson continued,

> Well, anyway, I did continue to work on the FAS [Federation of American Scientists]. But I also realized that I was no longer, that we were not effective. The FAS was not a very effective organization and especially when everybody in it was accused of being Communist. But the people who were at Los Alamos [working on weapons after the war] did continue to be very effective.

I asked Wilson if now he regretted his decision.

> No, but it was a personal thing, whereas previously it was a public thing. I suppose if I had tried and thought carefully about what would be the most effective thing to do as a good citizen, I might have continued to work, not necessarily at Los Alamos, but with the people at Los Alamos. I felt very strongly that we should not build a hydrogen bomb. Well, Hans felt that way too, but once the decision had been made to build it, then he was very effective in doing it, while I continued to say, "The hell with it." [The Russians] worked very effectively once they got going and, well, that may have led to a terrible war.
>
> But I was beyond all that; I was just a happy sort of a pacifist person from then on. Now, that's selfish in a sense. I mean, people who were working on it were strong anti-Communists, maybe they had very good reasons to be that way. And maybe Teller—he and I used to be sort of constantly at each other's throats—but he could have been more right than I was in terms of a big world fiasco.

Wilson was rethinking the decisions he had made after the war. While I had not assumed that he would be at peace with his life's moral choices, neither had I anticipated that he would be reconsidering them in this way. His actions did not achieve the results he expected they would. I heard the disappointment of a scientist who believed he should have had a greater impact in the social realm. He continued,

> I was as strong as I could be about not building the hydrogen bomb. Why should you build something that is a thousand times stronger than what we already have, which is too much? I wrote one of the first documents that came out of Los Alamos. I said that in five years the other nations would be able to build a bomb as well. I didn't know about Fuchs, of course.* The main secret was that the bomb could be built and any determined nation could do it, which turned out to be true. On the other hand, that the Soviets would get a hydrogen bomb and we didn't have a hydrogen bomb, well, what would that involve? Well, that was something that I guess I was prepared to accept. I wasn't a Communist or anything, but I was prepared to accept that kind of possibility. Which is hardly a good citizenly thing to do, I think.

As Wilson spoke, he smiled and sometimes broke into a good-natured laugh. He was not in the least egotistical or morose. Nor did he seem depressed by his self-evaluation. Yet I took what he said very seriously. He was considering the meanings of his difficult, carefully made choices and evaluating their logic and effectiveness. It seemed to me that he was still working them out for himself.

He recalled that from the outset he had been concerned about the potential impact of the atomic weapon on the United States' relationship with the Soviets. In fact, during the early Los Alamos days, he had discussed this issue with Robert Oppenheimer. Wilson reiterated that he had not been a Communist or anything close to it. In fact, he joked, as a young man from Wyoming he had barely known what Communism was. Wilson did believe, however, that the Soviets, as our allies, should have participated in the bomb-building project. He said that he

---

* The secrets that the German-born spy Klaus Fuchs passed to the Soviets shortened the time it took them to develop an atomic weapon. Hans Bethe estimated Fuchs's information may have saved the Soviets eighteen months. Soviet sources seem to confirm this. See David Holloway, *Stalin and the Bomb: The Soviet Union and Atomic Energy, 1939–1956* (New Haven: Yale University Press, 1994), 222, 417 n. 127.

often discussed this with Oppenheimer while they worked together organizing the laboratory.

> I saw a lot of Oppenheimer then, and the one thing that I do recall bugging him about was why it was that the British would come and work with us but the Russians wouldn't come. It seemed to me that down the line that was going to make some very hard feelings. And so he would never answer that at all—that wasn't any of our business. I don't know, I felt that perhaps he thought that I was testing him.

Wilson was alluding to the fact that even then Oppenheimer was under constant surveillance and therefore might have thought that Wilson was trying to set him up by asking if he did not think the Soviets should be involved. However, Wilson did feel strongly that the Soviets should have been a part of the project from the beginning.

> I would have liked to think that Roosevelt would have been able somehow to handle the Russians. I think he felt he could do that. On the other hand, he had Winston Churchill, who was very strong and was never clear on whether he was fighting the Germans or fighting the Russians. But it seems to me that by not involving the Russians and then assuming that they could do nothing about making a bomb, that was wrong.
>
> I mean they would have some reason for trust. But it was in fact that we didn't have them there and didn't tell them about it until it had been used. I don't see why they had any reason to trust us in that regard— I don't know. The mind boggles. Well, I didn't get anywhere with that [with Oppenheimer].

Thus we came to Wilson's relationship with Robert Oppenheimer. He remarked that he had not thought much of the theorist while at Berkeley. As a hardworking student of Ernest O. Lawrence, the inventor of the cyclotron, the young experimental physicist had not been a member of Oppenheimer's coterie of adoring students. Wilson told me, "I sort of disliked him. He was such a smart-aleck and didn't suffer fools gladly. And maybe I was one of the fools he hadn't suffered. But he was very sharp and I don't know . . . " Jane Wilson interjected, "Arrogant." Her husband agreed that "arrogant" was the correct word, adding that he had not cared for that aspect of the charismatic Oppenheimer's personality. "And," he recalled, "all of the theory students just adored him." Then, after a rocky start at Los Alamos, Wilson, although initially very critical of Oppenheimer's skills as a director, became one of his most loyal lieutenants.

> Very soon, in a matter of weeks, I could see that he was a good director. He was a quick study. He mastered all of those techniques right away. He was a person who was usually right, and I thought he was just a wonderful director. And I soon changed so that he would affect me, and when I was with him I was a larger person. Maybe that's why I was asking him those other questions, because *he* was so clever about them, I was probably trying to emulate him.
>
> I became very much of an Oppenheimer person and just idolized him. Everything that he had done was good and he was a good person to follow and a good person to talk to. I changed around completely. Then after the war was over I changed again.

Wilson's discussion of Oppenheimer brought me to one of my primary questions. I knew that when it became clear that the Germans were defeated, Wilson called a meeting in the Los Alamos cyclotron building. The subject was "The Impact of the Gadget on Civilization," a title he later saw as rather pretentious. ("Gadget" was a code name for the bomb.) In Boulder David Hawkins had told me that Wilson was the only person he knew at the laboratory who explicitly questioned the need to continue with the development of the bomb at that point. However, Hawkins said, Wilson had been convinced by Oppenheimer's argument that the atomic bomb must be made known to the world— that there was more at stake than the war in the Pacific. And, as Hawkins noted, it would have been hard for Wilson not to have been convinced. I now asked Wilson if that had indeed been the case.

> Well, yeah that's true. I thought we were fighting the Nazis, not the Japanese particularly. And that really, if I had been true to myself, then I should have left the project at that time—the time the Germans gave up. So I did. I called a meeting of the people to discuss it, and the meeting occurred. First, I asked Oppy about it because I couldn't do anything

without telling him. And he tried to talk me out of it, saying I would get into trouble with the G2, the security people.

As Wilson continued, I sensed the feisty defiance that lay beneath his pleasant demeanor.

So I said, "*All right. So what?*" I mean, if you're a good pacifist, then clearly you are not going to be worried about being thrown in jail or whatever they would do—have your salary reduced or horrible things like that. I guess I told him he hadn't talked me out of it. So then I went around and put notices all over the lab that there would be this meeting, and to my surprise he came to the meeting.

I asked if Wilson had known that Oppenheimer would attend the meeting.

No, I certainly did not. But there were staunch types like Viki Weisskopf [in attendance]. He [Oppenheimer] would have been in some trouble, I think, had he not come. You know, you're the director, a little bit like a general. Sometimes you have got to be in front of your troops, sometimes you've got to be in back of them. Anyway, he came and he had very cogent arguments that convinced me. Now I probably *wanted* to be convinced. But the main one was that he said there was going to be an organization meeting for the United Nations in April [1945], I think. And I think that this [meeting at Los Alamos] was, I don't know, maybe January. I remember there was snow all around.

And he, well, used this argument that the thing was to have demonstrated the bomb before that [U.N.] meeting occurred, because it should be organized on the basis, on the knowledge, that there are bombs. Whereas if we didn't show it, then the military people, their natural thing would be to keep it secret. And we were sure that was correct. So if we could test and show that it existed before that, then the United Nations would be set up at least on a basis of reality and not of nonreality.

And I thought that was a very good argument and it should be made on the basis of knowing that a bomb would either work or not work. And so it didn't have to do with whether it would be used or not used. It had to be something that *would be known* and that you couldn't keep secret. And that all of the nations who came there would know that this was a possibility, because of the test that we were to make before then.

So we should work. Of course we missed it. But we kept working even so, because still you knew there had to be a United Nations established at some time. And the United Nations should be set up on the basis of knowledge of nuclear bombs.

As Wilson pointed out, "they missed it." The inaugural meeting of the United Nations was held in San Francisco from April to June 1945, and the Trinity test was not conducted until July 16. But I told Wilson that I did not understand how, in Oppenheimer's mind, this secret military project would have been made known to the world. Even if the scientists had succeeded in making the test earlier, I wondered how the Los Alamos director imagined this knowledge would be brought to the U.N. Wilson answered that they could not be sure that it would succeed until the actual test and that they had to go to the United Nations with something proven.

> If you were setting up the United Nations, why should you set it up on the basis of something that wasn't true. I think he [Oppenheimer] had a great deal of faith in Roosevelt and I did too, and [in] the idea that they would do their best. This was a terrible problem, clearly—I mean an awful problem down through history—and here was an opportunity to nip it in the bud.

Wilson explained that he and his colleagues believed that sovereign nations could not continue to exist as they had, at least in terms of nuclear weapons.

> So there was an opportunity to handle the sovereignty problem, as the United Nations was set up. There would be areas in which there would be no sovereignty, the sovereignty would exist in the United Nations. It was to be the *end of war* as we knew it, and this was a promise that was made. That is why I could continue on that project. So I did. I called that meeting, he came to it, made this big pitch which I think was correct.

I recalled that Wilson first met Niels Bohr in 1936 and had been profoundly influenced by the great man's science and humanist philosophy. I knew that Bohr's vision included openness between nations on matters of scientific research and that he advocated international control of atomic weapons. However, I told Wilson that a more cynical view of what occurred at the meeting would be that Oppenheimer had used his argument simply to get Wilson and like-minded scientists to keep working on the bomb. The physicist replied, "My feeling about Oppenheimer was, at that time, that this is a man who is angelic, true, and honest and he could do no wrong. And I was interested in what he had to say and I believed in him." I wondered if General Groves, who was utterly devoted to secrecy and intent on using the bomb militarily, would have agreed with Oppenheimer's pitch to Wilson's gathering, had he heard it. Wilson responded, "No. But if you had any faith, that was the one thing that kept everybody together and kept all the Allies together—that there would be a United Nations made at the end of the war."

Wilson observed that another view would be that General Groves had chosen Oppenheimer to direct the bomb-building laboratory because he had something on him—namely, his Communist connections.

> Here is a man that he had right in his grasp, whereas the other scientists were independent types. Here was a person who wasn't independent at any moment. That's a good argument, I don't believe it, but it's still a good argument. Then Oppy could have made that pitch as part of his compact with the devil, who was the general. But I don't believe that.

I responded that I was thinking more in terms of power, that Groves had given Oppenheimer the power he wanted. Here, Jane Wilson, who had been quietly listening, reentered the conversation.

> I think Oppy loved being director of that project. I think he adored his fame as father of the bomb. And then it really changed his political and social outlook, so that from being an out-person, he became an in-person, and this of course happens to an awful lot of people. He was going to Washington, and this of course is now 1946 and later, until the fall [from power]. But he really enjoyed being a celebrity and a person in the halls of power. And I'm afraid we all would. So then, instead of feeling very much, the working man hasn't got a fair shake or we're going to do something for the teachers' union, we had bigger and better and higher things to do. And I'm almost certain that this went on. Don't you think so, Bob?

By "fall," Jane Wilson meant Oppenheimer's infamous "trial" before the Personnel Security Board of the U.S. Atomic Energy Commission. On May 27, 1954, Oppenheimer was stripped of his security clearance.

Wilson responded by saying that he had "always thought of Oppy as a person who had about five identity crises in his life." Then the serious subject matter took on a few moments of hilarity when Jane Wilson dryly added, "All because his hairstyle changed." Robert Wilson laughingly concurred, "You could tell from the length of his hair." No doubt responding to the incredulous look on my face, Jane Wilson elaborated.

> There he is the young radical professor at Berkeley—I was at Berkeley at the time too, though younger—and his hair was all little black curls. And then he was much more subdued at Los Alamos, the curls were not so curly. Then when he was a big wheel in Washington, he was quite shaved and quite sharp—quite mean, very arrogant. Really shaved and kind of looked like he was in the Marines, if you could be six foot tall and weigh about one hundred pounds and be in the Marines. Then, of course, after the fall, after he lost his position and was "defrocked" as it were, then his hair got gray and a little curly. And then we have the saint—he was very saintlike. He has kind things to say, he speaks sweetly and lovingly to all.

Wilson remarked to me, "Jane, as you might suspect, never gave herself up to Oppenheimer." And for her part, Jane Wilson explained that although she could enjoy being with him and acknowledged his importance to the project, she had never "felt he was real." When Wilson quickly added, "Anyway, I felt he was real," the conversation became serious again.

I told the Wilsons that whereas I had first thought of the Manhattan Project as the work of many talented individuals, I had recently begun to consider, perhaps with a more educated eye, the importance of Oppenheimer's leadership. I was learning that his person—his personality—had been a big force in the success of the project. Jane Wilson responded, "A big force, it was a big force. The fact that he was a good scientist was very important. If you are a good scientist, the scientists will forgive an awful lot, that's number one. And two, the enormous

respect people had for him and I did too." Wilson seemed to be growing uncomfortable with this discussion of his late friend and mentor
but added,

> He was a man that I knew that had charisma, that was competent. He
> had the same effect not only on the scientists there but everybody
> there. He used to talk about "the people's war" constantly, and I always
> thought that it was the people's war in many senses. I was convinced.
> Now I may have wanted to be convinced too. But that's human nature,
> so I may have just been looking for that. Certainly if you have been
> working on something for five years there's an innate curiosity, is it really
> going to work? You want to see it through. And somehow to be logical
> about it isn't very hard, that's human.

So Wilson had been convinced by Oppenheimer that the *development* of
the bomb should go forward. But had I understood correctly that this
did not mean he was agreeing that it should actually have been used?
He answered,

> I didn't agree. Of course, I always felt that it shouldn't be used. But I
> also understood that it was for the generals, for Stimson, for the presi-
> dent to make that decision, not for me. It's like joining the army—the
> general doesn't find out what your opinion is if you are a little private.

His reply led me to another issue I wanted to discuss: the betrayal he
had written he felt on learning of the bombing of Hiroshima. I had read
about this in his highly critical 1958 book review of Robert Jungk's story
of the Manhattan Project, *Brighter Than a Thousand Suns*. I had a copy of
the article and now asked Wilson about a particular passage:

> If Jungk is critical of the physicists who made the bomb, he is posi-
> tively censorious of the decision to drop the bomb on Hiroshima
> and Nagasaki. There are many who share this view; I count myself
> among them. It is only with his conclusion that I am in agreement,
> however, for here, as usual, Jungk oversimplifies and distorts his-
> tory. I recognize that my own feeling about the military use of the
> bomb was and remains largely sentimental rather than rational. *I
> felt betrayed when the bomb was exploded over Japan without discussion*

or some peaceful demonstration of its power to the Japanese [emphasis added].[2]

As I read the last sentence, Jane Wilson said, "Yes, he did. He was very upset." Her husband, who responded while she continued to speak, was more circumspect.

> Yes, but "betrayed"? I hoped it would be different. I may have felt betrayed, but that's like a private, you know. When the general does something that he doesn't like and he feels betrayed, but he knows he's going to be—I think I'm guilty of saying that incorrectly. When you are betrayed you can have a bargain with somebody, and we had no such bargain. I had zero bargain when I went into that. I knew that I didn't amount to anything—I mean that I was not powerful. The only thing that I ever felt betrayed about was the United Nations. I thought that this was used over and over to keep people fighting—that there would be a United Nations and this was going to be the last war. And I did feel that I was working on this project because of that bargain. Not that the bomb would be used or not be used, but that it would result in a United Nations.

While he examined the article he had written nearly forty years earlier, Jane Wilson continued to reflect on her husband's feelings about the bombings. She recalled that the announcement of Hiroshima came when she was away from Los Alamos, visiting her mother in San Francisco.

> I was very excited when the news broke in San Francisco. It all happened so fast that it kind of caught all of us off balance. But when I came off the train [in New Mexico] smiling and congratulatory Robert was very depressed, as a matter of fact. And all he needed was Nagasaki, which was going to happen almost immediately. He was really very unhappy, and there were these merry, impromptu celebrations at Los Alamos, where people were going around banging garbage can covers and so on, and he wouldn't join in, he was sulking and unhappy.

Wilson looked up from reading the article and said,

> Well, I think that the reason that I felt betrayed in some sense was that Oppenheimer had come to me. He had been appointed to a committee

about how the bomb should be used and where it should be used—this must have been in the last month of the war, I guess. And he asked me, he told me, to come in and talk to him about it. And he said that he was a member of the committee. And what was my feeling about it?

I asked Wilson if he was referring to Oppenheimer's membership on the Scientific Panel to the secretary of war's Interim Committee. He replied that he was and continued,

He gave me some time to think about it because obviously it was the biggest thing I've ever been asked to have an opinion about. And so I came back and said I felt that it should not be used, and that the Japanese should be alerted to it in some manner. And I said that we were just about to have the test in some weeks or a month and the Japanese could be invited to send over observers and they would be able to see this and they can go back.

We argued various parts, but he didn't believe that we should have people come over and observe. He thought that was not realistic, and it probably wasn't. They would have thought it was just some American scheme and they would not have sent people. Then we talked [about] another possibility, that they would be able to have some counter[measure]. For example, they knew something about radar, and they might have been able to set the bomb off at an earlier stage, which would have made it a dud. And then too high or too low—then it wouldn't have much of an effect compared to the effect it did have by being dropped at just the right altitude. Well, I didn't think much of that idea. Anyway, he said that he would think over what I'd said. And so maybe he did and maybe he didn't.

Anyway, obviously it had had no effect. But, on the other hand, I felt, and I might have felt betrayed because I had been arguing and he didn't tell me until it came out. Well, he shouldn't have. I mean it was a major secret. He should have had no business talking to me about it in the first place. But he clearly wanted some advice from somebody and he liked me, and I was very fond of him.

Since Wilson was questioning his past use of the word *betrayed,* I suggested that whether or not an actual betrayal had occurred, the word may have described his inner emotional state on learning that, contrary to his hopes, the bomb had actually been used. He began his answer by saying, "I felt bad about it, there's no question," and Jane Wilson interjected, "Oh, my dear." But Wilson continued,

> I now think I used the word *betrayal* there incorrectly. I guess I would have felt—the only person who could have betrayed me would have been Roosevelt. Because when Truman came on, well, he obviously had never even heard of it before, or Stimson might have. I had a great faith in Stimson. But I think *betrayal* is too strong a word and I was mistaken to use it.

With this, we came to a natural break in the conversation. It was midday, and Jane Wilson invited me to share a bowl of soup with them. As I helped with some of the preparations, she told me that she had not been well and was having trouble with her balance. While our meal was heating, I took the opportunity to look more closely at Robert Wilson's fine bronze, stone, and wood sculptures. I asked him about his career as a sculptor, and he explained that he had always been artistically inclined. In fact, during one sabbatical year, he studied at the Accademia di Belle Arti in Rome. He then told me that several years earlier he suffered a heart attack, during which he had no oxygen to his brain for a significant period. He had worked hard to recover his memory, but the left side of his brain was affected and he lost the ability to calculate. Thus his career as a physicist had ended.

Robert and Jane Wilson also told me that at one time they had planned to move into Kendal, Ithaca's newly built retirement community, with many of their friends and colleagues, including Hans and Rose Bethe. However, they reconsidered and decided to stay in their home, despite health problems. Wilson laughed out loud when I remarked that if I lived in such a wonderful house they would have to drag me from it kicking and screaming. A year after meeting the Wilsons I learned that, sadly, Robert Wilson had suffered brain damage from a massive stroke. It then became necessary for the Wilsons to move from their beloved home to the retirement community.

After the brief break, we had a little more time before Wilson needed to leave for an appointment at the university. I noticed that he seemed somewhat less focused and wondered if the morning's conversation had tired him. I resumed the interview by asking him about his experience at the Trinity test, about which he wrote,

> It was not until I directly observed the test explosion on the Jornada del Muerto desert of New Mexico that I had a real, an existential, understanding of what we had been making. . . .
>
> My first reaction, after being overwhelmed by its very existence, was to feel all my scientific and technical responsibilities literally just slide away: we had done our job. My second reaction, almost simultaneous with the first, was one of horror at what we had done, at what such a bomb could do.[3]

Now Wilson told me,

> It certainly was an epiphany. And whenever else would you have an epiphany except at seeing the first atomic bomb? From then on I certainly took myself more seriously. Up to that point I thought of myself as a physicist and what I could do was physics. Now I was concerned with what we could do, what a group of people, and what I in particular could do, about passing the word along and trying to have an effect to get people to take it [the bomb] seriously.

Looking back, did he see Niels Bohr as having a special wisdom and insight into the bomb's significance even before Trinity? "I certainly admired him tremendously. Well, one is inclined to ascribe all sorts of thoughts to people that you admire. Just as I had ascribed all kinds of things to Oppenheimer that probably he didn't quite live up to." Thus Wilson returned to the question of Robert Oppenheimer.

> Again, I feel that he had an identity crisis from the time he had been a professor to being the director of the lab. And trying to do his damnedest to be a good director, so that would be a second period. Then the period of sort of power, fight for power. He put himself forward to be the "Father of the Atomic Bomb." Really there were a lot of other people who had done much more with the project, like Fermi or

possibly Bethe, or Ernest Lawrence who had done a lot more than he
had. But somehow his personality just fit the press. I mean, whereas it's
always been a sort of a tradition not to put yourself forward. And here
was somebody who was putting himself forward and accepting that.
Oppenheimer accepted it happily and joyfully.

For his part, after the war Wilson threw himself into the organiza-
tional work of the newly formed Federation of American Scientists,
serving as its second chairman. I wanted to follow up on something he
had said about the FAS earlier in the interview—that the organization
had failed to effectively reach its goals. I considered the early work of
the atomic scientists' organizations and the FAS as crucial to the question of the social responsibility of scientists in the postwar era. Today it
is difficult to imagine how strange the idea of scientists having a voice
in public policy was in 1945. However, the newly exploded atomic
bomb was forging new links between what science makes and the social
world. The atomic scientists were among the first to understand this,
and many were deeply concerned about its meanings for a democratic
society and for international relations. I saw people like Wilson as pioneers in this new world. What did Wilson see as its, and his, failure?

He began his answer by saying the present organization is very
different from and much more successful than the one he helped found.
It was in the area of international control that the early FAS had, in
some measure, failed. When I commented that it was hard to imagine
how it could have succeeded at that time, he laughed and retorted,
"Yeah, but that's not succeeding." I agreed with him but added that it
was interesting to listen to him, because I had always thought the scientists had accomplished a great deal by organizing as they did—that they
had made a good, even noble effort. Jane Wilson was beginning to
speak as her husband said, "But it wasn't enough." She then answered
him,

> Well, yes and no, but you certainly became much more of a force.
> Now when there is a scientific question people still do appeal to the
> FAS. And it's an organization that's there that wasn't there and a little

bit more than that. I think, you consistently put forth your message and probably a lot of it stuck with a lot of people.

As Wilson recalled the long, dry period when they were essentially regarded as a "bunch of Communists," his wife continued, "Look at when Oppy went to see Truman. Truman said to his lieutenants, 'Don't let that crybaby in here again.' He didn't like him. When Niels Bohr went to see Churchill, Churchill said, 'Throw this guy out.' I mean—I think there are groups in society that don't communicate."

Jane Wilson was alluding to a meeting Oppenheimer had with Truman in the winter following the war, during which the president asked for the physicist's support in passing domestic atomic energy legislation. Oppenheimer suggested that it would be best to define the international problems first. He recalled saying, "I feel we have blood on our hands," and Truman responding, "Never mind. It'll all come out in the wash." The president later complained about Oppenheimer to Dean Acheson, calling him a " 'cry baby' scientist."[4]

Wilson answered his wife's point that some groups in society do not communicate by insisting that Bohr and Oppenheimer had nonetheless failed. "My understanding is that if you set out to do something and you don't do it, you fail. And we failed." Then for a few seconds Robert and Jane Wilson were speaking over each other, and I could not clearly hear what either was saying. On this question, he had the final word: "And we saved our conscience perhaps a little bit—and maybe had we worked harder. I don't think so, no, I don't think that we could have done anything else but give it that old college try, which we did."

We were approaching the end of the interview, so following on this, I asked Wilson how he envisioned the future. Was he hopeful? He said that there had been many opportunities to use atomic weapons, Korea or Vietnam, for example, or even during the last days of the Soviet Union, and they had not been used.

So I have a feeling that there is something in humanity, that makes us— there's good reason why we have been alive half a million years and that it has to do with continuing to do the thing that leaves you alive. That one of them is not to get over your head, and so, although there are lots of opportunities for bombs to be used, they were never used. And I have a feeling that there is something in humanity that makes them continue to exist. For that reason and for other shallow reasons I would say that I am an optimist.

So, *logically* I don't see any reason to be very optimistic. I don't mean about the next fifteen years, but in the next twenty-five years there seems to be every indication that we are only comfortable when we have a good enemy, and there's China out there just begging for it. So I could very well imagine that we have a situation which could only be resolved by us by the use of nuclear bombs. Well, but that's logic, and I've learned not to depend very much on logical arguments anymore. I'm an optimist because, just for the stupid reason I think there is a tremendous desire to survive.

I observed that he was *not* saying that humankind had learned important lessons over the last half century à la Niels Bohr's question about whether the bomb was big enough to bring the institution of war to an end. He was *not* saying that we had gleaned from the reality of nuclear weapons a new or deep understanding of the impossibility of war. Rather, he thought that our survival was a manifestation of something very old in human nature. Wilson answered, "That's what I'm saying." I pursued the issue by asking if he did not see an essential change in human thinking, if he did not see any lessons learned. He replied, "I don't think so. Only, except in an existential manner, I think we have learned. And I think that's why there has not been any interchange of bombs, because it is just too horrible to think about."

I closed the interview by asking him about his thoughts on the future of science as we move into the twenty-first century. I suggested that we would be faced with decisions about whether or not to pursue certain scientific research. Wilson replied,

I find it very hard to imagine that, because when an idea starts it's always such an innocent thing. Certainly when we started thinking about nuclei, it certainly didn't cross my mind it would be used in some horrible way. And I think it's the nature of science to sort of proceed crab-like, going from something that doesn't seem to be at all important to moving over a bit and it becomes very important. But if you try to go

directly into something and just discover something, well, that's engineering and not physics and not science.

Wilson told me that he did not know how we would be clever enough,
or have the foresight, to prevent scientific discovery that could lead
to unwanted applications. "Except," he said, "kill all scientists—that
might be a good idea. Throw them out in the snow as soon as they show
any curiosity. Let's get rid of them." Jane Wilson and I joined him in
joking about disposing of baby scientists, but there was something dead
serious in what he was saying.

> Well, that might be the way to go, for something stupid like that. It
> might be that people will realize the only way they are going to survive
> scientists or science is to get rid of the scientists. Anyway, I have a feel-
> ing that humanity will do the things that will cause it to continue to exist.
> That's very strong—we survived what we did.

The Wilsons both began to comment on all the horrors humankind
has survived: plagues, wars, famines. Robert Wilson added, "So we'll
probably continue, there's something built into us to realize what
should be done to survive. Well, this is a stupid thought, but that's my
thought." When I told him I did not think any of his thoughts were stupid, Wilson concluded,

> Well, not logical. I guess I would use Oppenheimer, one of the brightest
> people, certainly a scientist who just seemed so wrong time after time,
> I mean later on. And who had arrogant feelings about his ability, I think,
> to shape the future. Well, he couldn't shape it one bit.

October 29, 1996
Ithaca, New York

In the car on the way back to my hotel Robert Wilson and I continue to talk.
He tells me that Edward Teller was a friend at Los Alamos. Teller, the theo-
rist, was good and able at explaining physics to Wilson, the experimentalist.
He did not understand Teller's change in political views after the war.
*   Wilson says how much he respects the power of Hans Bethe's logic. He tells*
me that his own outlook tends to be less logical and more religious. Bethe, he

repeats, is one of his heroes for remaining relevant down through the years.
I feel an ache of sadness as he speaks and reply that I admire and respect the
strength of his deeply felt convictions. The complex web of individual
responses, when taken as a whole, might help us to comprehend the troubling
legacy of the bomb.

I tell Robert Wilson that in the larger design I believe his personal stand
against nuclear weapons development has been essential. And I do. *

Back at the hotel I had time to think about my day with Jane and
Robert Wilson. I recalled Wilson's description of Oppenheimer at the
"Impact of the Gadget" meeting, and it occurred to me that from the
start the Los Alamos director was a true maverick playing a dangerous
game. While he was running a laboratory for the military, he told the
scientists gathered at Wilson's meeting that they had to continue so
they would be able to thwart that same military in the coming fight for
secrecy and control. Oppenheimer directed a wartime laboratory
working toward a specific goal against an enemy. Yet he was simultaneously driven by his devotion to a vision of the universal meaning of the
mission—the end of all war. Apolitical in his early years, after
Hiroshima he actively pursued the influence that only the political
establishment could give him. The Wilsons spoke of Robert Oppenheimer's identity crises. Had his friends and followers mistaken his
potent intellect, broad understanding, and rhetorical skill for self-
knowledge? Had Oppenheimer, even with his great powers of concentration, confused himself to the point where he no longer knew his own
heart or mind?

I also thought about what Wilson said about the failure of the Federation of American Scientists to achieve international control and his
own inability to remain relevant to the nuclear debate. There was no
question in my mind that the atomic scientists accomplished a great
deal. However, the fact is, as Wilson made clear, international control
of nuclear weapons has not been achieved to this day. I can imagine the
deep regret that the scientists, my father included, must have felt when

* On January 16, 2000, while this book was in press, Robert R. Wilson died.

they lost what they believed was their only chance to prevent the development of the nuclear capability to destroy civilization.

For Wilson, the Trinity test was an epiphany. In that moment of revelation, what did the scientists see? Did some of them believe that their weapon had created a rift in time, a moment during which they could stand outside of history, beyond political cause and effect? Did they discern in the blinding light of the atomic blast exactly where to step through in order to change humanity's course and end all war? Looking back, it seems clear that this was not in their power. Had they seen something real, or had their perception been altered by the bomb's overwhelming brightness—were they thrown off center by the shock wave of their creation? Perhaps the scientists were not unlike their territorial forebears, the Spanish conquistadors, who four centuries earlier saw gold in the sun-reflecting walls of the Southwest pueblos, and who, in their quest for El Dorado, and their own escape from history, tried to cut across the unknown territory of La Jornada del Muerto only to die of thirst.

✳ Professor Bethe at Home in His Office

The day after my visit with the Wilsons I went to Cornell to see Hans
Bethe. I found him sitting at his desk, which was covered with papers
and books. Three of the four walls had bookshelves floor to ceiling; a
window in the fourth overlooked the campus. On the blackboard was a
simple drawing that appeared to be two elliptical orbits with direc-
tional arrows, around a center point. During his talk earlier in the week,
Bethe had briefly discussed the Nazi bomb project. Of particular interest to me was the extent to which Werner Heisenberg and other German scientists had or had not been loyal to the Nazi cause and the
impact of this on their bomb research and development. How much the
Germans actually understood about making an atomic bomb as well as
the reasons for their failure to develop it remain subjects of intense historical debate. Interestingly, one way in which the issue of the morality
of scientists working on weapons of mass destruction has been framed
is the comparison between the Manhattan Project scientists and their
colleagues who remained in Germany during the war.

In *Brighter Than a Thousand Suns*, Robert Jungk asserted that German
physicists, including Heisenberg and Carl Friedrich von Weizsäcker,
refused to build an atomic bomb on moral grounds, consciously preventing Hitler from obtaining the powerful weapon. His corollary argument was that Allied scientists, by working on nuclear weapons, failed
to uphold the true spirit of the scientific tradition. In 1958 Hans Bethe,
like his colleague Robert Wilson, critically reviewed the book, saying
that as far as he could tell ethical considerations played only a minor
role in the German scientists' decisions. Bethe reported that Heisenberg told him that he had estimated it was far beyond Germany's capacity, and therefore that of the United States, to develop an atomic
weapon during the war.[1]

Now, forty years later, I wondered how Bethe saw the question. At
the time of our meeting I was reading Ruth Lewin Sime's biography of

the physicist Lise Meitner. The discovery of fission is another controversy arising from the relationship between scientific development and war making during World War II. In 1944 Meitner's German colleague, radiochemist Otto Hahn, alone received the Nobel Prize in chemistry for the discovery of fission. Sime offered a compelling analysis of the relationship between Meitner and Hahn and his suppression of Meitner's leading role in the discovery. According to Sime, such falsification and distortion of reality were central to Nazi Germany, and this legacy persisted even after its defeat: "[Hahn] suppressed and denied not only his hidden collaboration with a 'non-Aryan' in exile but the value of nearly everything she had done before as well. It was self-deception, brought on by fear. Hahn's dishonesty distorted the record of this discovery and almost cost Lise Meitner her place in its history."[2]

Sime argued that such perversion of history and memory also occurred among the ten leading German scientists who were interned at Farm Hall, a country estate near Cambridge, England, after the defeat of Germany. While they were incarcerated in comfort, their conversations were recorded through hidden microphones. During their internment, the scientists received the news of Hiroshima. It was then, Sime asserted, that the Germans, eager both to explain their failure to achieve the atomic weapon and to justify having stayed in Nazi Germany, created the legend of the German scientists' moral superiority.[3]

Within hours of learning about Hiroshima, Weizsäcker stated, "I believe we didn't do it because all the physicists didn't want to do it, on principle. If we had all wanted Germany to win the war we would have succeeded." He went on to say, "History will record that the Americans and the English made a bomb, and that at the same time the Germans, under the Hitler regime, produced a workable machine [critical reactor]. In other words, the peaceful development of the uranium machine was made in Germany under the Hitler regime, whereas the Americans and the English developed this ghastly weapon of war."[4]

In her commentary on Weizsäcker's proposition, Sime stated, "By implying *could have* along with *would not*, it turned German scientific shortcomings into evidence of moral scruples: now the Americans were the mass murderers and the German slate, at least with respect to fission, was wiped clean." Confirming one of Bethe's original criticisms, Sime noted, "Recently Jungk has stated that he was too willing to believe Heisenberg and Weizsäcker, because at a time of McCarthyism in the United States he was anxious to show that scientists could resist the power of the state." On the other side of the issue, Thomas Powers asserted, "If Heisenberg can be said to have refused, in whatever degree, to build a bomb for Hitler, then Allied scientists are in effect invited to explain why they were justified in building a bomb for Roosevelt."[5]

When I asked Bethe about Heisenberg and about Sime's analysis of the German scientists' Farm Hall mythmaking, he replied,

> Well, only Weizsäcker did so, not Heisenberg. And I have studied the Farm Hall report rather carefully, at least the part that is relevant. It seems very clear, that the Germans did not actually try to make a bomb. But the reasons for that are a different matter, and I think Heisenberg said it correctly when he said, "We just couldn't. Anything we might have done, to separate isotopes would have been a very big enterprise which we simply could not afford with all the other pressures of the war. And also if we had built a big factory, it would have been bombed." That is one side of the question.
>
> The other side is Heisenberg gave involuntary but direct evidence that he didn't try to build a bomb because when [after learning of Hiroshima] his colleagues asked him to talk about the critical mass, he presented something which made no sense at all. What he produced in his first [Farm Hall] talk was patterned after the calculations you do to make a reactor, which he knew very well. And which are totally inappropriate for building a bomb, because in a reactor you have to slow down the neutrons—you have a long diffusion length from the creation of the neutron to its absorption. And that's what Heisenberg used.
>
> For a bomb, on the other hand, the neutron which you produce at one point is directly used to make the next fission. It took him about a week to catch on to that, and a week later he gave a very proper and good talk about the correct theory, in which he did essentially what [British bomb project scientist Rudolf] Peierls and Frisch had done in 1940. Farm Hall conversations, more than anything else, proved the point that the Germans really did not try to make a bomb. So, now, an additional point, if the Germans had tried to make a bomb, they

certainly would have tried very urgently to measure the fast neutron cross-section of uranium 235. And once they conquered Paris, they had a cyclotron which was eminently suitable for the purpose. And Heisenberg never requested any measurements of this kind, nor did anybody else. So I think it is well established by the Farm Hall records that the Germans did not try to make a bomb. Heisenberg, I think, gave the correct explanation, namely, that they realized it was beyond their capacity.

We then turned to a question of weapons and morality that was closer to home. I told Bethe that although we had discussed it before, I still found it difficult to understand why he had decided to work on the hydrogen bomb. The previous day Robert Wilson had provided another interpretation of Bethe's choice. I told Bethe that I had reread his 1950 piece against the development of the hydrogen bomb, published during the heated debate regarding the crash program to develop the weapon. Robert Oppenheimer headed the General Advisory Committee to the Atomic Energy Commission that advised against the development of the H-bomb. However, Edward Teller, along with the State Department, argued in its favor, and President Truman, afraid that the Soviets would get there first, ordered its development. Although Bethe was not a member of the GAC, he was consulted during its deliberations. In 1950, after the decision to proceed had been made, he argued:

> I believe the most important question is the moral one: Can we who have always insisted on morality and human decency between nations as well as inside our own country, introduce this weapon of total annihilation into the world? The usual argument . . . is that we are fighting against a country which denies all human values we cherish, and that any weapon, however terrible, must be used to prevent that country and its creed from dominating the world. It is argued that it would be better for us to lose our lives than our liberty, and with this view I personally agree. But I believe this is not the choice facing us here; I believe that in a war fought with hydrogen bombs we would lose not only many lives but all our liberties and human values as well. . . .
>
> We believe in personal liberty and human dignity, the value and importance of the individual. . . . We believe in peace based on

mutual trust. Shall we achieve it by using hydrogen bombs? Shall we convince the Russians of the value of the individual by killing millions of them? If we fight a war and win it with H-bombs, what history will remember is not the ideals we were fighting for but the methods we used to accomplish them.[6]

Nonetheless, Bethe did eventually consult on the development of its design. In a previous conversation he had told me that he would have been happy if the scientists had remained ignorant of how to build the hydrogen bomb. And I had read the reminiscences of Bethe's close friend, Victor Weisskopf, who also opposed its development. He described a long discussion between the two men during a drive from Princeton to New York in October 1949. At the time Weisskopf believed that he had convinced his friend to refuse Teller's request to work on the thermonuclear weapon. He recollected,

> Unfortunately, two events in early 1951 influenced the future of the H-bomb: First, Teller and Stanislaw Ulam found a relatively simple way to make the weapon. This made it a virtual certainty that the Soviets would also soon be able to construct it. Second, the Korean War broke out. . . . Under the influence of these two events, many physicists who had previously expressed doubts about the H-bomb began to participate in its development. Among them was Hans Bethe.[7]

Now I wanted to better understand Bethe's reasoning. I asked if it was correct that both the scientific advances and the political situation had influenced his decision. He replied,

> The project of the hydrogen bomb really was limping along. And nobody knew how to do it until [mathematician] Stanislaw Ulam had the idea that what one had to do was to compress the hydrogen to a much higher density and that the way to do this was by means of an ordinary atomic bomb. Anyway, Ulam went to [then head of the Theoretical Division at Los Alamos] Carson Mark and said, "I have this idea and I'm afraid this may make the hydrogen bomb possible." Ulam didn't want the hydrogen bomb. So Carson Mark told him, "I'm sorry you have to tell Teller." So he did, and Teller immediately took hold of that idea, he improved the manner in which the A-bomb was used to compress the

hydrogen, and they wrote a joint paper. And then Teller pursued it together with an assistant, [physicist Frederic] de Hoffman, who was a good friend of mine. And they wrote a second paper in which they outlined how one could actually put that in practice. Now I was told about both papers, I didn't get the papers, I was here [at Cornell]—somehow, I got to know the details.

When I asked Bethe if these details had not been secret, he said that they were "very, very secret." So Bethe was very well informed on the progress of the hydrogen bomb. Once he realized it was possible, he also knew that if the Americans could do it, so could the Russians.

And indeed, so they did. Then once this was clear I thought that it should be done. And so then I went to Los Alamos on Teller's urging, in '52, and oh, I guess for at least half a year, I kept in constant contact with the hydrogen bomb development—there was a weekly meeting. I didn't contribute terribly much, but my main contribution came when Teller, who was very hostile to Los Alamos, wrote us, "It won't work because of the following reasons." And so I sat down and examined that and concluded that Teller was wrong. That was my chief contribution.

I asked Bethe if, even while doing this work, he had remained worried about the implications of the thermonuclear weapon.

I still considered it a mistake and a calamity, but since it obviously was possible to do it, and since we had orders to do it, and the Russians presumably would discover it too, there was nothing to be done but to do it. Truman had given the order to do it.

Did that mean that his original plea against developing the H-bomb had been addressed to the decision makers, not to the general public?

Yes. That is right. However, it was too late, because, well, I made a plea verbally. I made a plea to Oppenheimer, who was all of my opinion, but then came Truman's decision and the published plea which was made together with a dozen other physicists at the meeting of the American Physical Society, in New York, in January of '50. At that time Truman had already decided. And so, we knew essentially that our plea was in vain and that Los Alamos was committed by the government to do it. But fortunately nobody knew how to. Early '51, when Ulam made his invention, Teller immediately improved on it.

I asked if, in his opinion, the push to prevent its invention was doomed from the outset, because even if the United States had not developed it, the Russians might have done so anyway. Bethe answered, "Exactly so. In fact, in [Russian physicist Andrei] Sakharov's memoirs, he states that if we had said 'We'll refrain from doing this,' the Russians would have considered this a deception—they just want to mislead us and want to persuade us not to do it." I wondered if this did not actually make Teller's argument that it had to be done.

It was very difficult for Truman to decide otherwise. It would have been perfectly safe, however, for Truman to say, "We will do the research, and stop short of the test." That in fact was proposed by Vannevar Bush—and that would have been perfectly safe, because if we did, you can't build a hydrogen bomb without [a] test. And so if we had said that and somebody else had tested, well, we would have tested. And that would have been the right thing to do. But Bush got nowhere, as you can imagine.

So Bethe was telling me that he went along with the work even though he was not happy with it. "That's right. I was doing it; there was nothing I could do to prevent it." I asked if Wilson was correct in saying that, by staying with the work on the weapon, Bethe had thus been able to have a voice in government decision making on nuclear policy.

Well, I stayed with it, and so I could participate in the President's Science Advisory Committee, which couldn't prevent the hydrogen bomb, that was already past history, but we prevented for a long time the antimissile. Unfortunately, we had our eyes firmly fixed on the antimissile and we didn't see that something was going on which was more dangerous yet. Namely, multiple warheads. And we were actually told about it, but we didn't come out against it. And I think that was a great omission, and other people did come out against antimissiles. But we were in a position to do so, and having our narrow vision on the antiballistic missiles, we just paid no attention to the multiple warheads.

We spent more time talking about nuclear weapons and arms control, and then I told him about my travels to see his colleagues. But I also wanted to speak to Bethe about his reflections on his life in physics. I asked him if he had a sense of himself as a historical figure in twentieth-century science or in the entire history of science. He replied, "I think so. I'm not an Einstein or a Bohr, I'm not even a Heisenberg, but somewhere I am part of the history of science." What did he mean when he said he was not an Einstein or even a Heisenberg? "Well, their imagination, their penetration into the unknown." I asked if, when he spoke to people, as in his lecture earlier in the week, he *sensed* himself as a historical figure. He said, "Yes," and laughed as he added, "As my son said after the talk, 'Well, they got it from the horse's mouth. And there aren't so many horses left.' "

I had a final question for the day. I told Bethe that I knew people consider the twentieth century a great era of scientific discovery. As a physicist, did he think that he and his colleagues had left the world better than they found it?

That's a difficult question. We have contributed to the world a lot of good things, like radar which won the Second World War, like lasers which certainly are very useful, like safety of airplanes in flight due to radar. Certainly life has become more comfortable due to science. We have also contributed evil things like the atomic bomb, which I considered unavoidable once the discoveries were made.

I don't know whether we left the world better, we certainly left it more comfortable. But anyway, that's *applications* of science, not *science itself*. And one thing that I think ought to be made very strongly, science is very different from the applications. The concepts of science, the fact that you can actually prove things, that's a very different matter from all the applications that I quoted.

Science is finding how elementary particles work, and how nuclei are made, and how an atom works and makes chemistry, and how chemistry makes biology. That's science. Application of science, this is a different matter.

I responded that scientific knowledge itself is generated with or without applications. And, I said, looking at where we stand at the end of this magnificent century of physics, the question arises, in what ways has this knowledge enriched or impoverished our welfare?

Well, that's what I have discussed. I think, when you talk about pure knowledge, pure science, not atomic bombs, not hydrogen bombs, not nerve gas, then I agree with Teller that research should be unlimited. That is, one should just try to understand as much as one can. And we may come to a limit, where we just can't continue. But Rabi, who is a great hero of mine, has always made this point very strongly that I made a minute ago, namely, science is different from its application. And people generally don't understand what science is about.

I commented that many criticisms of science by very smart people were often criticisms of technology. He answered, "They usually talk about technology. But one cannot deny that technology is made on the basis of scientific knowledge; but it is totally different."

CHAPTER 6

Joseph Rotblat,
Pugwash Pioneer

In October 1995, while listening to the radio, I learned that Joseph Rotblat and the Pugwash Conferences on Science and World Affairs had received the Nobel Peace Prize. At the time of the announcement, I was writing a letter of introduction to Rotblat, Pugwash president, who was a member of the British mission to the Manhattan Project. Reports of the Nobel Prize stressed that most people had no idea of what "Pugwash" was, but the organization was well known to me. I have vivid memories of my father's excitement about his participation in Pugwash during the 1950s and 1960s.

In 1990, when I was caring for my father after my mother's death, he was attempting to reconnect with a career that, although over, still held meaning for him. Sometimes he would ask me for the address of a colleague he had lost contact with. I was particularly touched by the brief correspondence he initiated with the Russian physicist I. M. Franck, whom he had not seen for many years. The two men wrote across time, and culture, to tell each other of their losses and of the isolation brought by failing health. In his last letter to my father, Franck proposed, "As it is not easy for us both to meet because of our health, let us keep in contact with the help of letters. I think our friendly feeling to each other remains unchanged."

One day I came upon my dad in his easy chair reading the proceedings of a recent Pugwash conference that a friend had sent. Although he had not been to a meeting since 1967, the question of the proper role of the scientist in society was still very much with him. It was a poignant moment for me when he said he wanted to attend the upcoming conference. Physically, my father had great difficulty traveling. His mental condition was deteriorating, and he lacked awareness of the severity of his illness. I felt like the bad guy as I explained to him that I just did not think it was possible. I knew that my reasons did not stand up to his desire to act on his deeply felt convictions. My dad, in his childlike way, was unable to understand my logic—he wanted to go, he should be able to go. At moments like these I was most aware of the loss he suffered when his career and, to a great extent, his way of life had been cut short by his illness. And I could not help but feel that I was, in a very real way, failing him.

After my father died a Brookhaven colleague, and close family friend, sent me his papers from the lab. Looking through the documents and photographs, I found a file on Pugwash that contained my dad's correspondence with Rotblat. The Pugwash Conferences on Science and World Affairs were an outgrowth of a 1955 document known as the "Russell-Einstein Manifesto." In December 1954 Bertrand Russell gave a radio addressed titled "Man's Peril," which warned of the dangers should nuclear weapons be used in another war. From this he drafted a manifesto calling for the abolition of war, as well as for a conference of scientists to work on the task of international cooperation and understanding. Russell sought endorsements from scientists of varying political views. Among the first he approached was Albert Einstein, who signed the document two days before his death. Thus the name "Russell-Einstein Manifesto." It stated,

> In the tragic situation which confronts humanity, we feel that scientists should assemble in conference to appraise the perils that have arisen as a result of the development of weapons of mass destruction. . . .
>
> Here, then, is the problem which we present to you, stark and dreadful and inescapable: Shall we put an end to the human race; or shall mankind renounce war? People will not face this alternative because it is so difficult to abolish war. . . .

There lies before us, if we choose, continual progress in happiness, knowledge, and wisdom. Shall we, instead, choose death, because we cannot forget our quarrels? We appeal, as human beings, to human beings: Remember your humanity, and forget the rest. If you can do so, the way lies open to a new Paradise; if you cannot, there lies before you the risk of universal death.[1]

With the financial support of the American millionaire and philanthropist Cyrus Eaton, twenty scientists from ten countries met in Eaton's birthplace, Pugwash, Nova Scotia, in July 1957. Thirty-eight years later Pugwash and Joseph Rotblat were awarded the Nobel Peace Prize. The Nobel committee recognized Pugwash's work in bringing together scientists and decision makers, in spite of political differences, to collaborate for the reduction of the nuclear threat. The committee hoped that the Peace Prize would encourage the world's great powers to strengthen their efforts to rid the earth of nuclear weapons.

I was prepared to go to London to interview Professor Rotblat, but I also told his office that if he came to the United States, I would travel anywhere in the country to see him. We eventually arranged to meet in Chicago, in March 1997. Although I had been to the city several times over the years, I found myself looking forward to this trip in a different way. Not only was I going to interview Joseph Rotblat, but I would be seeing my parents' childhood home with new eyes.

As I prepared to leave, I realized that the journey was taking on the quality of a return to the Old Country. I found myself trying to visualize my mother and father growing up during the depression, children of immigrants who had maintained many customs of their homelands. I imagined my mother as a teenager, hurrying down the street to her ballet class. I pictured my father, studying his engineering texts late into the night in his family's cramped apartment near Armitage and Central Park. And I saw him riding the streetcars and trains to Northwestern University, on the shores of Lake Michigan.

I arrived in Chicago a few days before my interview with Rotblat, to spend some time with my cousins and to explore the city. One night, after a long walk in the sharp March air, I lay in bed, straining to discern through the static of the bedside clock radio the slow strings of a Schubert quartet. Huddling under the covers, I realized that on some level I had expected to actually physically reconnect with my parents, to see them, or at least to concretely grasp something of their youth. Instead I traversed the chilly streets alone, feeling that they were just ahead of me—vanishing into an alley before I could reach them, carried around a corner by a gust of wind, oblivious to their daughter's pursuit. I was in the right place, but my timing was out of phase.

On the morning of the interview, I reread my notes. Joseph Rotblat was born in Warsaw in 1908. A research fellowship brought him to the University of Liverpool in April 1939. There he worked with the discoverer of the neutron, James Chadwick. Rotblat returned briefly to Poland in August and had been back in England only a few days when Hitler invaded his homeland. His wife and family perished in the Holocaust. Rotblat's work with Chadwick led to his membership on the British bomb team, and to the Manhattan Project—Chadwick headed Britain's scientific mission to Los Alamos. When it became clear that the Germans had never possessed a viable bomb project, Rotblat decided to leave Los Alamos and did so at the end of 1944. After the war, concerned about how his physics research could be applied, he went into the field of medical physics. For more than twenty-five years he was a professor of physics at the University of London and chief physicist at St. Bartholomew's Hospital. Rotblat served as secretary general of Pugwash from 1957 to 1973 and as president from 1988 to 1997, when he became president emeritus. In 1998 he was made Knight Commander of the Order of St. Michael and St. George.

Late in the afternoon I took the elevator to Rotblat's floor and knocked on his door. Rotblat himself answered. I had assumed that a man of his age would be traveling with an assistant; however, when we met I discovered he was alone. He was full of energy, fit and spry for his near-ninety years. His speech was rapid-fire, the traces of a Polish accent overlaid with a British veneer. The room was spacious, with a sitting area well suited for my recording equipment. Before formally beginning the interview, I showed him a snapshot of my father standing with three other men. I knew it had been taken at a

Pugwash conference, perhaps one in the late 1950s. I had tried to identify my dad's companions from Pugwash records and wanted to confirm my guesses with Rotblat. He examined the picture and told me that the three gentleman in suits and ties were K. S. Krishnan of India, G. P. Thompson of Great Britain, and I. Ogawa of Japan. The fourth man, my father, for reasons that must remain mysterious, was wearing Bermuda shorts, a T-shirt and suede jacket, knee-high socks, and a beret!

In preparing for the interview, I had been thinking about the experiences of the scientists like Teller, Bethe, and Fermi who had fled Nazism, soon to find themselves in crucial relationships with their adopted countries. I was just beginning to appreciate that they had been real strangers in their new homes and at the same time entrusted with considerable power. I opened the interview by telling Rotblat that I was particularly interested in the experiences of the refugee scientists, but he immediately and emphatically corrected me.

> I am *not* a refugee. People often mistake me for a refugee. I left Poland because I received a research fellowship. It started in April 1939 and was supposed to last one year. But after a few months, the war broke out, and the political situation changed completely and I couldn't go back—I wanted to go back, but I couldn't. And eventually the situation changed again in Poland, so I decided not to go back after the war. But I was not a refugee. *I did not run away.*

I was feeling nervous and vulnerable, as I often do during the opening moments of an interview, and silently wondered why my mistake had elicited such a strong reaction. Yet I knew that Rotblat had left Poland on a fellowship and had been unable to return after Hitler's invasion. Aware that my question had been imprecise, I resolved to be more careful. I asked Rotblat to tell me more about what had originally brought him to Chadwick and England.

> I had, in fact, two invitations to do research, one was from Chadwick in Liverpool and the other was from Joliot-Curie in Paris. I always say that any sensible person who's got a choice between Liverpool and Paris would choose Paris, obviously. I did not—I was not sensible. But it turned out to be a fortunate decision. If I had chosen Paris, I wouldn't

Figure 14. My father with K. S. Krishnan, G. P. Thompson, and I. Ogawa at a Pugwash conference, Kitzbühel, Austria, 1958.

be alive now. There was also a good reason why I decided to go to this dreadful place, Liverpool, with its slums: it was my intention to raise the status of physics in Poland, and I wanted to build a cyclotron in Poland, and at that time Chadwick was building a cyclotron in Liverpool.

I had read with interest Rotblat's 1985 article about leaving the atomic bomb project. In it he reported that he had learned of the discovery of fission while still in Poland, and also realized that this meant an atomic explosion might be possible. He wrote that the idea of using his knowledge to produce such a weapon was abhorrent to him. However, his work with Chadwick led to his participation on the bomb project. I asked if he could tell me more about why he had used the phrase "my scruples were finally overcome" when describing his decision to work on the bomb.[2]

Yes. Well, I have always thought about science in humanitarian terms. Many scientists think about science for its own sake—just discovering the facts of nature. This is all they want to do science for. For some reason, from the very beginning, I felt that science should have a purpose. The purpose is to serve mankind. I have always followed Francis Bacon's thoughts in *Instauratio Magna*, that science should be not just for the pleasure of the mind but also to improve the lot of man. To work on the atom bomb was completely against these principles. I mean, a scientist, in my opinion, should never work on a weapon, let alone a weapon of mass destruction. And these were my scruples.

This was the reason why I had this terrible time during the summer of '39, because I knew the war was coming and I knew that Hitler was going to invade Poland. Everybody in Poland knew it was going to happen—it was only a matter of when and not if. But I was afraid, really afraid, that German scientists may make the bomb and that this would enable Hitler to win the war. The danger of this happening—of Nazism taking over the world—this I couldn't stand. This was enough to convince me to overcome my scruples about being involved in a weapon of mass destruction.

I told Rotblat that I was interested to learn he had been conscious of going against his own principles. Was his reasoning in favor of the bomb that he considered the defeat of the Nazis a higher good? He quickly answered, "Against my own principles, that's right," but with a sharp laugh, he corrected me. "The immediate danger, this is important. I wouldn't call it a higher good."

I asked him what kinds of things he had been thinking about as a young man, when he conceived of atomic energy as a benefit to mankind.

Atomic energy could, can be a benefit. It has two faces, good and evil. I mean, the good part is that one can use atomic energy to generate electricity. There are some problems, sure, but nevertheless energy is needed. That's only one part. There are many other applications of the discovery. I've used it myself a great deal in medicine, using radioactive materials for diagnosis and for therapy. There are many good aspects to it, no doubt about it. But there are all the bad things, the evil things, as well.

Aware of the passion with which some people oppose all uses of nuclear energy, I wanted to make sure I understood his views. I was posing my question as a statement, "So, you are a proponent of nuclear energy," when Rotblat broke in, insisting that I not jump to this conclusion.

> Initially, and I must say even after the war, when the bombs were used against Hiroshima and Nagasaki, I was at that time a proponent of nuclear energy. To a large extent it was more psychological than real; namely, I was concerned that science had acquired a bad image in the minds of people. People learned about nuclear energy for the first time, not from reactors but from bombs. They learned there's a new source of energy which can destroy. And if everybody began to think of science as something bad, I as a scientist would feel this is wrong. Therefore I felt we should somehow make up for it and show the good side of it. I must say that I somewhat exaggerated the good side because I wanted people to change the image they held of science. So when you ask if I am a proponent, I should say, "Maybe, yes, at one time I was." But since then I've found that our views at the time—that atomic energy is cheap, clean, and safe—were wishful thinking. And none of this is completely true. Nevertheless, it is an important source of energy— many countries depend on nuclear energy now. We can't ask France to shut down its reactors, it would immediately shut down France. We've got problems to solve, the disposal of nuclear waste, for example, we don't know how to do it yet.
>
> But the main thing that worries me is in connection with nuclear weapons—that the materials coming out of reactors can be used in nuclear weapons. And therefore I feel that in the long run we should not rely on nuclear energy, we should put in an effort to find other sources of energy. I hope I have now clarified my position.

Saying I was glad he had, I went on to ask if he thought there was ever a chance, with the discovery of fission coming so close to the beginning of World War II, that the evil side of atomic energy, the bomb, would not have been developed in one country or another. He responded,

> I think even during the war it was still possible to prevent this from happening if we followed Niels Bohr's proposals. You know, his idea was

that the Allies, America and Britain, should share the secret with the
Russians, on the condition they would agree to an international control.
That was the chance.

I knew of Rotblat's opposition to the atomic attacks on Hiroshima
and Nagasaki and wondered if he thought an agreement with the Soviets during the war would have directly influenced the decision to use
the bomb. "Well," he replied, "if the Soviets had become involved, then
the chances of Japan surrendering would have been greater. As it was,
Japan wanted to surrender before the bomb was used."

Rotblat, like so many of his Manhattan Project colleagues, judged
Bohr's farsighted ideas of the bomb's wider implications to be of great
significance. But did this mean that Bohr was against the actual bombings? As far as I could tell, it did not.

However, the Metallurgical Lab's Franck Report did explicitly link a
demonstration of the bomb, as opposed to its military use during the
war, with postwar international control. A few months earlier I had met
with the discoverer of plutonium, Glenn Seaborg, the last surviving sig-
natory to the Franck Report. He said that he did not ruminate or worry
after the bomb was used, because it had brought an abrupt end to the
war and had saved many lives. Nevertheless, Seaborg told me, the
Franck committee's goals were dealt a blow by the dropping of the
bomb—looking at the question from the point of view of the arms race,
it would have been better if the weapon had never been used in warfare.
That way, Seaborg said, the United States would have been in a better
moral position.

My understanding was that Bohr saw the wartime and postwar
issues as distinct. Bohr's primary concern, and the focus of his wartime
efforts, was that an agreement for international control of atomic
energy between the United States, Britain, and the Soviet Union be in
place at war's end. He came to Los Alamos asking Oppenheimer, "Is it
big enough?" To me, the question implied that Bohr believed, with
Oppenheimer, that the world would have to experience the bomb's full
force to comprehend its meaning. I asked if Rotblat thought Bohr was
against using the bomb during the war. He did not answer directly but

made reference to the question of whether we know Bohr's views on this issue.[3] "I don't think that he was actually consulted after the failure of his 1944 proposals. He was very disappointed—very unhappy, because he knew what was going to happen. Afterwards, he came back to this idea of openness. This was his main idea."

In his article about leaving the project, Rotblat reported a Los Alamos incident that had shocked him and had raised doubts about the ultimate purpose of the scientists' work. At the time Rotblat was living with the Chadwicks. He recalled that the Manhattan Project comman-der, General Leslie Groves, frequently dined at the Chadwick home when at the laboratory. Rotblat described one such private occasion, when Groves said that, of course, the real purpose in making the atomic bomb was to subdue the Russians. While not claiming to quote his exact words, Rotblat wrote that Groves's meaning had been clear. "I felt deeply the sense of betrayal of an ally. . . . Until then I had thought that our work was to prevent a Nazi victory, and now I was told that the weapon we were preparing was intended for use against the people who were making extreme sacrifices for that very aim."[4]

Aware that some historians argue that the bombs were used on Japan, not to bring about its surrender, but to make the Soviets more manageable after the war, I asked Rotblat if he thought that Groves had been giving voice to official policy. His immediate answer was that he would rather not say, as he did not know. When I stated that I was sim-ply curious about what he thought, he replied, "I know that some mili-tary people thought in the same way, or he wouldn't have said this. He repeated this ten years later, you know, in public." Rotblat told me that he had been accused of lying when he related what General Groves had said in his presence. But, he asserted, his vindication came ten years later, in the form of the Atomic Energy Commission's 1954 hearing on Robert Oppenheimer's security clearance. Rotblat paraphrased General Groves's official testimony—that within two weeks of taking over the bomb project, he had conducted it on the basis that Russia was the main enemy. He suggested I look this up.

On returning home, I read Groves's testimony at the Oppenheimer trial. In the section Rotblat had paraphrased, Groves stated, "I think it is well known—that there was never from about 2 weeks from the time I took charge of this project any illusion on my part but that Russia was our enemy and that the project was conducted on that basis. I didn't go along with the attitude of the country as a whole that Russia was a gallant ally. I always had suspicions and the project was conducted on that basis."[5]

I told Rotblat that I could not help but think of the historical controversies surrounding the reasons for the use of the bomb on the Japanese cities. He replied that he was not a historian but only knew what he had heard Groves say, before the bomb was made or used.

> Whatever may have been the motivation for the use of the bomb on the Japanese cities, to me this was definitely the first step in the nuclear arms race, no doubt about it. So whether intended or not, it resulted in the nuclear arms race and this is what Niels Bohr predicted.

Rotblat was invoking Niels Bohr and had alluded to his disciple, Robert Oppenheimer. This brought to mind something else I wanted to ask about his article on leaving Los Alamos. In it Rotblat not only discussed his own scruples but also analyzed the moral consciences of project scientists in general.

> When it became evident, toward the end of 1944, that the Germans had abandoned their bomb project, the whole purpose of my being in Los Alamos ceased to be, and I asked for permission to leave and return to Britain.
>
> Why did other scientists not make the same decision? Obviously, one would not expect General Groves to wind up the project as soon as Germany was defeated, but there were many scientists for whom the German factor was the main motivation. Why did they not quit when this factor ceased to be?
>
> . . . The most frequent reason given was pure and simple scientific curiosity—the strong urge to find out whether the theoretical calculations and predictions would come true. These scientists felt that only after the test at Alamogordo should they enter into the debate about the use of the bomb.
>
> Others were prepared to put the matter off even longer, persuaded by the argument that many American lives would be saved if

the bomb brought a rapid end to the war with Japan. Only when peace was restored would they take a hand in efforts to ensure the bomb would not be used again.

Still others, while agreeing that the project should have been stopped when the German factor ceased to operate, were not willing to take an individual stand because they feared it would adversely affect their career.

The groups I have just described—scientists with a social conscience—were a minority in the scientific community. The majority were not bothered by moral scruples; they were quite content to leave it to others to decide how their work would be used.[6]

Then, in the very next paragraph, Rotblat referred to a wartime proposal involving Oppenheimer that had never gone anywhere—the idea of using radioactive strontium to poison the German food supply. I asked Rotblat if he had purposefully placed this discussion of Oppenheimer directly after his assertion that most project scientists lacked moral scruples. Was he intentionally making a statement about Oppenheimer's morals and motivations?

Well, I'm afraid I may be harsh on Oppenheimer, but this is in fact what I intended. I had a great admiration for Oppenheimer at the time that I met him in Los Alamos. Really, he was a hero for me, for various reasons, including that I believed him to be a humanitarian. Gradually things came to my knowledge. I felt, this is not the way a hero of mine should behave. Gradually he became an antihero.

For example, the fact that he agreed that the bomb should be used on the cities. He could have said no. And at that time he was powerful enough that his voice might have prevailed. He agreed, together with the military, that it should be used against civilians. This was one of the things which I felt was wrong. Later on came the thing which brought him down—he opposed the hydrogen bomb. I always thought that he opposed it for moral reasons, the same way that Rabi did, or Fermi did, on pure moral grounds. And then I found out he did not oppose it on moral grounds, at least, this is, not openly. He said that, militarily, there was no need for America to have the hydrogen bomb—we'll do much better by developing tactical nuclear weapons. And again, I was greatly disappointed that he should have taken this stand.

The final disappointment came when an American historian sent me a copy of the letter which Oppenheimer wrote to Fermi in 1943. Now I happened to be very much involved with radioactivity, and the effects of radiation. And it struck me, how can a person of such standing, like Oppenheimer, even think about poisoning food that would be eaten by women and children?[7] How can anyone ever sink to such depths?

The only way I can explain it is by two factors: one is, such power corrupts and Oppenheimer knew power. He became to a certain extent overwhelmed with power. And the second, a more general one, is the effect which war has on moral standards. As soon as war breaks out our moral standards break down. And I think this probably happened with Oppenheimer. At that stage, he became part of the war operators, and this is the reason why he felt, with the generals, that the bomb had to be used against Japanese cities. This is the reason why I am so much against war. It could create conditions when I would once again behave in the same way. It *could* happen again. I'm a pacifist, I always say, but I'm not an absolute pacifist.

This brought me back to his statement about overcoming his own scruples. He was not an absolute pacifist, so, in the case of his bomb work, there had been real reasons for his scruples to be overcome—the threat seemed great enough.

Exactly, this is what I say. This is something, a weakness perhaps, in people. In order to overcome my scruples I had to develop a rationale—I had to live with it—namely, that we needed the bomb in order to prevent Hitler from using his. In other words, I developed the concept of nuclear deterrence. And therefore, the reason why I am not an absolute pacifist is because I still live in the real world. People may think I am a dreamer, but I'm a dreamer with my eyes open. I still have to find some reasons for behavior, and I can't exclude that in certain circumstances I would do a similar thing again.

As so often happened during the interviews, I found myself trying to comprehend the dreadful circumstances of World War II. I was remarking that the fear of Hitler must have been tremendous, especially among the Europeans, refugee or not, when Rotblat asserted, "Well, the evil. The concept of Nazism is an absolute evil. If there is any absolute, there it was."

Rotblat, like Oppenheimer, credited conversations with Niels Bohr as contributing to his consciousness of the social and political implications of nuclear energy and nuclear weapons—that the bomb held a deeper meaning demanding international understanding. Yet their interpretations of Bohr's insights had spurred them to sharply contrasting actions. Freeman Dyson, who has written about both men, believes that Oppenheimer's tragic flaw was his restlessness: "Restlessness drove him to his supreme achievement, the fulfillment of the mission of Los Alamos, without pause for rest or reflection." The scientists working under Oppenheimer, he argued, were caught up in his restless pace. In contrast, Dyson wrote, "Only one man paused. The one who paused was Joseph Rotblat from Liverpool, who, to his everlasting credit, resigned his position at Los Alamos . . . when it became known that the German uranium project had not progressed far enough to make the manufacture of bombs a serious possibility."[8]

Implicit in Dyson's assessment is the notion that there was an essential shift in the bomb project's meaning—a shift that few stopped long enough to perceive, let alone act on. I asked Rotblat if he actually experienced a pause, or if it had always been clear to him what he would do in such a case.

> No. It was always quite clear to me. Of course, I had to pause, I had to take a stand. But it was something which I would have anticipated taking. Because if there is a reason for doing something, and this reason turns out to be invalid, then the logical step is to proceed a certain way. It's pure logic. My rationale [for starting the work on the bomb], as I mentioned before, was to prevent Hitler from using his bomb. From the beginning the intention was that the bomb should not be used at all. When the cause of my worry ceased to exist, it was quite logical that I should not proceed any further.

I proposed that someone could have joined the project, as Rotblat had, with the conception of the weapon as a defense against the Nazi

nuclear threat, and then, as the situation evolved and new factors
emerged, such a person might actually change his mind. The prospect
of the invasion of Japan, for example, might cause him to believe that
the bomb should be used. Rotblat replied,

> If I had changed my mind, then I would have also to change the reason,
> the initial reason. If there was any worry that the Japanese may have
> nuclear weapons, then, of course, this would still apply, as it applied to
> Germany. But I never had any reason to believe that the Japanese were
> working on it. And with what we knew about the state of Japanese sci-
> ence at that time, it did not seem likely—they couldn't make it.

I explained that I was not referring to the question of the Japanese having a bomb but to arguments that the bomb brought a swift end to the
war. Rotblat was emphatic. "No, but it was using the bomb that I was
completely against. My whole purpose, in doing the work, was that it
should never be used, not even against Hitler. I never contemplated that
we would use it."

I was curious about what Rotblat knew of Leo Szilard, because my
reading was that he, like Rotblat, had conceived of the bomb as defensive—as a deterrent to the Nazi nuclear threat. He responded,

> Well, he would have preferred that the bomb was not used at all. But if
> it was to be used, it should be used as a demonstration on some unin-
> habited island. This was his idea. In the second petition he brought out
> this argument about not using it against people. My argument would
> have been, stop the work altogether, there's no need to continue.

Rotblat has written that in 1944, when he told Chadwick that he
wished to leave Los Alamos, he discovered that the project's intelligence arm had assembled a thick dossier on him—he was suspected of
being a spy. He was known to have visited someone in Santa Fe without
official permission. The theory was that he had arranged with this contact to return to England, then to be parachuted into Soviet-occupied
Poland, in order to give away atomic bomb secrets. Rotblat had justified
his unauthorized activities in the article, and I was intrigued by the way
he characterized them: with Chadwick's permission, he had been visit-

ing someone in Santa Fe for purely altruistic purposes. Rotblat had provided no further details, and this piqued my curiosity. So I asked if he
would say with whom he had been meeting, assuring him that I would
respect any wish not to tell me. He replied,

> No, no, I can tell you. It's a long story. But I was vulnerable—I was
> vulnerable indeed. It's a very moving story really. People may look at it
> in the wrong way, they wouldn't believe what I say. But I have been
> absolutely honest about this. I once wrote about it in a piece entitled
> "The Spy That Never Was."

When I responded that I had not read it, he explained that it had never
been published. Our conversation about the implications of the atomic
bomb then took a long, winding detour. I sat quietly as Rotblat recalled
a half-century-old story of coincidence and friendship unfolding on the
edges of the Manhattan Project.

Shortly before leaving England for Los Alamos, while at a party, Rotblat was introduced by a mutual friend to a young woman, a law student at the University of Liverpool. Her mother was British and her
father American. She had inherited a congenital hearing defect from
her father; her nerves were atrophying and she was going deaf. Soon
after meeting the young woman, he left for the United States, having no
idea of his specific destination. He had met her only once.

Then, while in Los Alamos, he received a letter from the friend who
had introduced them at the party. It turned out that the young woman
had come to the United States. At the age of twenty-one she decided to
become an American citizen, in order, perhaps, to get the best medical
treatment. The doctors could do nothing for her but suggested that she
move to Santa Fe for the clean air. Their mutual friend, not knowing
where Rotblat was, asked that he at least write to the young woman, as
she was very lonely. Telling me the story, Rotblat marveled at this accident of history and again asserted that it would be difficult to believe.

As Santa Fe was the only place project scientists were allowed to go,
he decided it would be nice to visit the young woman. However, before
doing so, he was required to inform the project's intelligence officer.
Rotblat said that he had always been a rebel about such things and did

not want to report his personal comings and goings to the authorities. At the same time, because of his loyalty to the British team, he asked Chadwick, who gave him permission to visit the young woman and told him that it was not necessary to notify the intelligence officer. He visited the young woman once and did not intend to return.

However, about a month later, after D day, Rotblat received a letter from the young woman informing him that her brother had been one of the first killed in the invasion. Knowing she was in a terrible state, he went to Chadwick, and again received permission to visit her. Then other things happened—her German shepherd was requisitioned by the army, and she had no way of knowing when someone was at the door. So he hooked up an electrical system, so that her lights would go on whenever the doors opened. When Chadwick left Los Alamos to live in Washington, D.C., Rotblat simply continued to visit the young woman on his own from time to time. He took up flying lessons in Santa Fe on Sundays and would stop by her apartment before going back to Los Alamos. Rotblat assured me that the friendship was completely innocent. Nevertheless, based on these activities, and the fabrication of others, when he asked to resign from the project, army intelligence made the case that he was a spy. When he was able to disprove the false allegations, he was given permission to leave.

As Rotblat explained in his article, the chief of intelligence instructed him not to discuss his reason for leaving the project. An agreement was made with Chadwick that his departure would be described as purely personal: he was worried about his wife in Poland. A second condition of his leaving was that he have no contact with his project colleagues. Thus, Rotblat reported, the bombing of Hiroshima took him completely by surprise: "I was devastated. Since the defeat of Japan was already obvious, I saw the destruction of Hiroshima as a wanton, barbaric act, and it made me very angry."[9]

As we returned to the discussion of his ideas, I wanted to better understand the relationship between his views of the scientists' roles during World War II and their roles during the cold war. I knew that he passed harsh judgment on the Manhattan Project and its outcomes. In

*Figure 15. Joseph Rotblat,
Santa Fe, 1944.*

his 1995 Nobel lecture he stated that, with Hiroshima,
"a splendid achievement of science and technology had turned malign.
Science became identified with death and destruction. It is painful to
me to admit that this depiction of science was deserved."[10]

Following on this, did Rotblat really believe that the scientist, not
the commander in the field, was at the heart of the arms race? He
responded,

> In a way, the scientists took over the lead role from the military people.
> The momentum of the arms race was determined by the scientists, not
> by the military. And this is very bad. Again it shows how you get yourself
> involved in a certain way and forget that you are a human being. It
> becomes an addiction and you just go on for the sake of producing a
> gadget, without thinking about the consequences. And then, having
> done this, you find some justification for having produced it. Not the
> other way around.

I wondered if it was not also a question of developments occurring within a historical context. I suggested that the argument could be made that scientists worked on the development of nuclear weapons, especially once it was known that the Russians would have them, because there was now a possibility of aggression from them. Did not the whole concept of deterrence come from the idea that there had to be a balance between the two powers? I suggested that even though the scientists were responsible on one level, they worked within a larger frame—world politics were going on at the same time. Rotblat asserted that this was not the case.

> I'm sorry, I don't agree with you. Suppose we start off on the concept of deterrence, although I disagree with it nowadays, as you know, but suppose we agree that America needed a deterrent against a Soviet attack with nuclear weapons. How many bombs would be needed for this purpose? [Former Secretary of Defense Robert] McNamara judged that about four hundred would be enough for deterrence. Why would one need any more? But this is what happened, America built up a huge arsenal. And, of course, the Soviets followed suit. Then the scientists said, "Well, let us now find ways in which we can destroy the Soviet missiles before they can be used." And, of course, the Soviets did the same. It's complete madness. This is what is worrying me, because, you see, at the beginning even though the bomb was terribly destructive, and we knew that the hydrogen bomb could destroy the largest city in the world, still you could not destroy the human race. But with the gradual growth of the nuclear arsenals, it came to a stage when we could destroy the whole human race. This could happen with a hundred thousand of them. And this is why I never could forgive the scientists for getting themselves into such a mad arms race.

Listening to Rotblat, I was uneasy. He seemed to have interpreted my questions to mean that I condoned or supported the arms race, or the maintenance of a large deterrent force, which I do not. I was actually trying to get a sense of how he saw the relationship and interplay between the scientists' actions and the larger societal and historical forces. I looked down at my notes and heard him remind me that we had to finish up before 6:00 P.M. Taking a deep breath, I moved on. There was more I wanted to ask.

Rotblat and other Pugwash participants have worked hard over the years to make the ideals of the Russell-Einstein Manifesto operational in the real world. As an advocate of the total global elimination of nuclear weapons, in 1995 he testified before the International Court of Justice on the illegality of nuclear weapons. He advocates a system of societal verification. Once an international treaty banning nuclear weapons is achieved, citizens, especially scientists, would have the legal obligation to report to an international body on illicit nuclear weapons programs in their countries.

I told Rotblat that I had been reading his writings on societal verification and whistle-blowing—the notion that individuals have to exercise their consciences by putting humanity above nationalist concerns. I wanted him to discuss his views on dissidents and on spies. He has written about Klaus Fuchs, the German-born member of the British bomb team who successfully spied for the Soviet Union while at Los Alamos. When he met with Fuchs a few years before his death, he found him to be an ardent Communist.

> [Fuchs] was convinced until the end that he performed a noble deed. In a way, one can find a justification for Fuchs' actions: he guessed that the main purpose of the Manhattan Project was to give the United States overwhelming military superiority over the Soviet Union—I can testify from a personal occurrence that this was largely true—and he decided to restore the balance.
>
> Looked at from this angle, he could be considered as a dissident. Yet, in my opinion, there is a big difference between spying and whistle-blowing. . . . What Fuchs did was to transmit information in secret to a regime notorious for its suppression of freedom of information. Many scientists on the Manhattan Project believed that the Soviet Union should be invited to participate in the control of the development of nuclear energy in both its peaceful and military applications. Foremost among them was Niels Bohr. . . . He advocated the sharing of knowledge, but did not try to contact directly Soviet authorities.[11]

I prefaced my question by saying that it was my understanding that Rotblat did not quite agree with Fuchs, because he acted in secret. He emphatically responded, "I didn't agree with him at all!" I continued by saying that my question also had to do with something Rotblat had written about the Rosenbergs. He told me,

> As far as I know, they were both of them spies. I don't know enough of the history of the Rosenbergs, but I know quite a bit about the history of Fuchs. In fact, I talked to him, that's why I can speak more about him than the others. So, in the case of Fuchs, I definitely felt this is not the way to fulfill your whistle-blowing [societal responsibility]. If you feel that there is something going on which the general public should know about, then you take the risk, as Vanunu did.* Say it in the open and do not convey the information to a regime which is even worse than the one which you are betraying. This is what I felt about Fuchs. Whistle-blowing means telling the public, and not telling something in private— this is spying. And I would never accept spying.

Rotblat had clarified his views for me, as my question had been exactly about the spies acting in secret, which is fundamentally different from the humanitarian openness advocated by Niels Bohr. I showed him the paragraph he had written about the Rosenbergs that had so puzzled me:

> At present, loyalty to one's nation is supreme, generally overriding the loyalty to any of the subgroups. Patriotism is the dogma; "my country right or wrong," the motto. And in case these slogans are not obeyed, loyalty is enforced by codes of national criminal laws. Any transgression is punished by force of the law: attempts by individuals to exercise their conscience by putting humanitarian needs above those dictated by national laws are denounced by labelling

---

* Mordechai Vanunu is a former Israeli nuclear technician who, in 1986, gave information about Israel's secret nuclear weapons program to *The Sunday Times* of London. Seized in Rome by Israeli agents, he was convicted of treason and sentenced to eighteen years in solitary confinement in an Israeli prison. After eleven and a half years he was released from solitary confinement. He has been denied parole, and his contact with the outside world remains severely restricted. Vanunu denies ever being a spy, insisting that he is a whistle-blower motivated by humanitarian concerns.

those individuals as dissidents, traitors or spies. They are often severely punished by exile (Sakharov), long-term prison sentences (Vanunu), or even execution (the Rosenbergs).[12]

After reading it he said, "Yes, well, here I speak about *transgression*, you see, and transgression could be either coming out in the open, or could be spying." He explained that in this particular case, he was not making a point about a specific type of activity—whistle-blowing or spying. Rather, he was highlighting the consequences of such transgressions of national loyalty—that they could "even come to a death sentence." So he had not meant that there was an equivalence between the acts of Sakharov, Vanunu, and the Rosenbergs? He replied, "No, no, no."

Rotblat is dedicated to the notion that we must develop an allegiance to humanity that transcends national boundaries. And he places the responsibility for nuclear weapons and their consequences directly on the scientists' shoulders. This, in combination with the conviction that science should serve humanity, is at the heart of his activism. It all hearkens back to the thesis of the Russell-Einstein Manifesto—that nuclear weapons make war impossible, that by realizing this, and abolishing all forms of war, humans will reach their true potential. Challenges to national sovereignty, although difficult, must eventually lead to a day when we will not be torn apart by such loyalties. In a nuclear age, we cannot afford to be.

Rotblat wrote that a new mind-set is needed before the idea of a war-free world can be accepted universally: "Just as in the distant past our main concern was for the safety of our family, and just as later this extended to the security of our nation, we must now begin to be conscious about the protection of humanity. In the course of history we have gradually been enlarging our loyalty to ever larger groups. We must now take the final step and develop an allegiance to humanity."[13] When I told Rotblat that I was interested in learning how he thought we could achieve an allegiance to humanity, he replied, "Actually, I don't know myself. I'm still probing. I'm still looking for it."

At the time of the interview, I happened to be reading a book related to Rotblat's concerns. *For Love of Country* opens with Martha C. Nussbaum's essay arguing for the cosmopolitan view. She invoked the Stoic notion of conceiving of ourselves as being surrounded by a series of concentric circles, from the self out to family and community and finally to humanity as a whole. Nussbaum asserted that while affirming our children's love for family and community, we must teach them the values of world citizenship.[14]

The remaining essays in the book are responses and critiques, some in support of and some opposed to Nussbaum's thesis. The discussions bring to light the complexity underlying the phrase "citizen of the world." Of particular interest to me was Sissela Bok's careful examination of the metaphor of widening concentric circles of human concern and allegiance—a metaphor much like Rotblat's concept of "enlarging our loyalty to ever larger groups." Bok wrote,

> The metaphor . . . of concentric circles of human concern and allegiance . . . speaks to the necessary tensions between what we owe to insiders and outsiders of the many interlocking groups in which we find ourselves. It is a metaphor long used to urge us to stretch our concern outward from the narrowest personal confines towards the needs of outsiders, strangers, all of humanity, and sometimes also of animals. . . . But more often it has been invoked to convey a contrasting view: that of "my station and its duties," according to which our allegiances depend on our situation and role in life and cannot be overridden by obligations to humanity at large. . . .
>
> Henry Sidgwick took the contrast between the two perspectives to be so serious as to threaten any coherent view of ethics. . . .
>
> Both the universalist and bounded view concern human survival and security. I agree with Sidgwick that neither can be dismissed out of hand as morally irrelevant.[15]

One of my questions about moral issues facing nation-states versus a world government has to do with obligations to one's citizens versus obligations to the world. Choices are often made in the face of real dilemmas. Conflicts may be based on deeply held national, cultural, and individual morality. I had been thinking about the case of war, and how the issue of the decision to use the atomic bomb might be looked at in this light. I put the question to Rotblat: suppose we accepted, for the moment, the disputed assertion that the atomic bomb saved American lives. Are leaders morally obligated to save the lives of soldiers over the lives of the enemy? Or do they have an obligation to humanity never to put civilians in harm's way? When I said that it seemed to me that real conflicts exist, not just in war, he responded,

> Of course, there's conflict—but the problem won't be there if there's no war. My objective is to create a world without war. Therefore, this conflict couldn't arise. I mean, of course, it exists now and this is terrible for individuals like myself, for example, who have to live with divided loyalties.

He emphasized that as long as war exists, difficult personal choices will remain. The danger is that now, with nuclear weapons, the escalation of war could destroy us all.

When I told Rotblat that I was still thinking about the problem of the bomb's use in World War II and not nuclear war in today's world, he answered,

> Okay, therefore I say in between, we will always be subject to these conflicts. I can't give a general formula for the way to go. It depends on the circumstances. When you speak about the interest of the nation, very often it is not of the nation but of the government that happens to be in power. You will probably find that another government would have approached the problem in a different way.

It was time for me to close the interview, and I had some final questions. Freeman Dyson had recently told me that we no longer have much to fear from physics, which he believes has already done its worst. It is biology that he finds both troubling and exciting. I wondered if Rotblat was concerned about current developments in the biological sciences and saw a role for himself and Pugwash in the debate. He replied,

> Very much concerned, yes. Because I keep saying that we know that humanity can be destroyed, humans can be destroyed with nuclear weapons. But as science goes on uncontrolled, then some other means of total destruction, perhaps more readily available than nuclear

weapons, could be developed. And I can think about genetic engineering. This is a field which has got such great potential, which one wouldn't have believed until a few years ago. And you see these stories about cloning and so on, which means to me, people go on doing experiments without thinking where they are going to lead.

My response would be to set up an international ethical committee in which projects which may have important consequences would not be allowed to go on until they have been vetted by the ethical committee. Take the example of clinical research which involves patients. Nowadays most of us just cannot go on doing these sorts of experiments as research. Each project has to go to an ethical committee in the hospital. If it doesn't fulfill certain conditions, then the project won't be permitted.

People would say, "You put controls on research." Well, sometimes you may have to put controls on research. Coming back to what I said at the beginning, science should be for the benefit of mankind, not just for its own sake. I come back to my basic philosophy about this. This is why I say now we have to do this on an international basis.

I asked whether it would be the scientists themselves or their governments who make such agreements. He said emphatically, "No, no. I keep saying it should *not* be the government but the scientists themselves. Pugwash is largely for this purpose, to get the scientists to be conscious of their social responsibilities."

Did he think that young scientists were sufficiently aware of these issues?

Not enough. But more than they used to be, because at one time scientists really lived in the Ivory Tower. The *outcome* of their work was not their concern. I have a feeling that, particularly in the United States, the young people are now more conscious of the consequences of their work than before.

I wondered if, as the century came to an end, Rotblat thought there were lessons we are in danger of forgetting.

On the whole, I believe that every century is better than the previous century. This century has been one of the worst, true, but at the same time, in spite of all the terrible tragedies and millions of lives lost in various wars, the standard of living is going up, people live longer everywhere. Slavery is gone completely. So, from this point of view, we can see gradual progress—as the evolution has to be. From the beginning things have been evolving, and, of course, it is never smooth, you know, it goes up and down. But, overall, we are going in the right direction.

And therefore we should now look forward to the twenty-first century as getting further along on the road of enlightenment, with more time for cultural achievements in the world, in the arts, and literature, and in relations between people. The Internet, you see, is one very big step forward. Because it is very important that people should be able to talk to each other. Ignorance is one of the big obstacles. Ignorance is the fomenter of strife and war.

I remarked that it sounded as if, even in our postmodern world, Rotblat held to Enlightenment values. He replied, "I do, yes." In closing, I asked Rotblat how he wanted people to think of him. Was how he would be remembered important to him?

I don't think that it is important—any single person is not important. I don't believe in the cult of the individual. I feel that if I have managed to achieve something, this is my satisfaction. So certainly, I'm glad. I'd be most disappointed to say that my life had passed without having achieved something. But the main thing is that each of us is a member of the community. The important thing is what all of us together have achieved rather than the individual.

✳ The Old Country

Joseph Rotblat is the only scientist from Poland whom I interviewed. On the plane ride home I reflected on his impassioned rejection of a nationalism that demands allegiance "overriding the loyalty to any of the subgroups." Rotblat has lived through a century of vicious wars between nation-states. He cannot afford the luxury of being pessimistic about the future. In 1939 he left his homeland on a scientific fellowship, and within six years the world of the Polish Jews had vanished. I wondered if I had discovered a clue to Rotblat's intense reaction to my mistaking him for a refugee. "I did not run away," he said. He had not abandoned his wife and family to the Nazis.

My father's mother, Sarah, was born in Poland, the granddaughter of a rabbi. The great sorrow of her life was that she was unable to bring her young niece, the daughter of her eldest sister, Libbe, from the old country to Chicago before Hitler's invasion. One branch of the family escaped to South America. Libbe and her daughter died at the hands of human beings transformed into monsters at the distorted extreme of national and "racial" loyalty. My father's cousin searched diligently for thirty years, looking for any trace of the family. There was none to be found. A long time after the fact it seemed fitting that I had met Joseph Rotblat in Chicago.

*Figure 16. My great-aunt Libbe (left) and her family,
Poland, 1938.*

CHAPTER 7

Herbert F. York,
Inside History

I had the opportunity to visit Herbert York at his La Jolla home on several occasions. He and Sybil, his wife of fifty years, live in a comfortable rambling house magnificently situated above the rocks and tide pools of the southern California coast. York was the first director of the Lawrence Livermore National Laboratory and the first chancellor of the University of California at San Diego. His government service has included membership on the President's Science Advisory Committee and the General Advisory Committee on Arms Control. During the Carter administration, he was the ambassador to the Comprehensive Test Ban negotiations in Geneva.

Now retired, York is professor and director emeritus at the University of California's Institute for Global Conflict and Cooperation, which he founded in 1983. He also serves on the University of California President's Council on the Nuclear Laboratories. During World War II, the University of California managed the Los Alamos laboratory for the federal government. The tradition has persisted, and today, under contract with the Department of Energy, the university manages the Berkeley, Los Alamos, and Livermore labs.

Like many other youthful project members, including my parents, there was nothing in York's background to indicate that he would be working on the top-secret Manhattan Project with the country's leading scientific lights. In his case, of primary importance was Ernest O.

Lawrence, inventor of the cyclotron, winner of the 1939 Nobel Prize in physics, and founder and director of Berkeley's Radiation Laboratory.

Herbert York was born in Rochester, on Thanksgiving Day 1921, into an old upstate New York family. His father, who never completed high school, was a railway shipping clerk, like his father before him. His mother had been a secretary at the Eastman Kodak Company. York wrote that his early school days were not promising. However, he apparently underwent a metamorphosis in his final two years of high school and performed so well that he received a scholarship to the University of Rochester. He described his university experience:

> The first year in college provided some of the greatest thrills of my life. I still remember vividly the discovery of all manner of things I had previously been totally unaware of: the joy of learning how the world actually worked, the existence of such things as graduate students, Ph.D. degrees, and, above all, mature people who did what they were doing because they enjoyed it and not just to earn money to support their families. College, I found, meant a great deal more than simply not working on the railroad, and I was determined to become part of this newly discovered world. . . . I met many very special people there whom I would otherwise never have encountered.[1]

Among his professors at Rochester was the eminent refugee physicist Victor Weisskopf. York told me that the cultured Austrian was the first "world-class" person he had ever encountered. "I was terribly impressed by him in terms of what he was, what he knew, a physicist who did physics and knew all the great ones, had been in Copenhagen [with Bohr]." Weisskopf had fled Hitler, traveling a circuitous route through France, England, Switzerland, and Denmark before arriving in Rochester. He left Rochester for the bomb project before York did and was a key group leader in the Los Alamos Theoretical Division, headed by Hans Bethe.

It would take great imagination to invent the strange and unique set of historical circumstances that would place Weisskopf and York concurrently at a small upstate New York university and then further connect them as scientists building the first weapon of mass destruction. And it was even more unlikely that they would be brought together thirty-five years later as members of a select group of experts advising Pope John Paul II on a speech to UNESCO addressing the danger of nuclear weapons and the evil of nuclear war. But all this, and more, came to pass.

York was just shy of his twenty-first birthday when a project recruiter from Berkeley sought him out at the recommendation of a former Rochester physics professor. In May 1943, master's degree in hand, York left New York for the University of California to work with Lawrence and his cyclotrons. From Berkeley, he went to the Manhattan Project Site Y-12 in Oak Ridge, Tennessee, where he was a member of the scientific team producing enriched uranium. York recalled the culmination of his work in Oak Ridge.

> One day in the late spring of 1945, word came that we were to shut down all Calutrons on a certain June date and take all the material in the collectors out for processing and shipment to Site Y (Los Alamos). Normally, once a Calutron received its input charge, it took about a month of continuous operation to process all of it; only then were its product collectors removed and their contents processed for shipment. To stop in the middle of a "run" was inefficient. This sudden change from the norm could, therefore, mean only one thing: the project was coming to its culmination. . . . The magic date came. We took all the product we then had, processed it and sent it off.[2]

The word "Calutron" was coined from California University Cyclotron. Lawrence created the huge machine, which separated uranium 235 from uranium 238, on the principles of the cyclotron and the mass spectrometer. What York and his team shipped to Los Alamos was the costly uranium 235, fuel for Little Boy, the Hiroshima bomb.

For young Manhattan Project scientists like York and my parents, the wartime bomb-building effort was an opportunity to work with many of the scientific giants of the era. Some recruits were still students, often the first in their families to attend college. And they were working with the living legends of their field, European émigrés like

Fermi, Szilard, Bethe, Teller, and Weisskopf and illustrious Americans like Oppenheimer and Nobel laureates Lawrence and A. H. Compton. I came to understand that more than an atomic bomb was built during the war years. Whether they intended it or not, the scientists were forging powerful bonds between their academic research, government, the military, and the universities. York's own career is a significant link in this chain of connections.

Several times over a period of three years, I rode the train two hundred miles down the California coast to meet with Herb York. He is a friendly, gregarious man with a razor-sharp mind, and I have come to feel at ease speaking with him. While he holds strong opinions, he is able to see many sides to an argument and is intellectually tolerant of views that differ from his own. Our dialogue is ongoing, and whenever we sit together in his living room, I have the sense of picking up the thread of conversation from my previous visit.

One afternoon, without a specific question to start with, I found myself thinking out loud: Robert Wilson had told me he was always against the bombings. On leaving Los Alamos after the war, he gave up his security clearance, vowing never to do such work again. But, looking back, Wilson had said that by refusing to do weapons work, he believed he had effectively removed himself from the nuclear debate about which he cared so deeply. As I tried to formulate my question, York jumped in.

It depends on *how* repulsed you were by Hiroshima. It may be wrong, and maybe it's just a much later rationalization, but in my own case, I simply say there were fifty million people killed in the war and very few at Hiroshima that ended it. So I don't have any feeling of regret at all except in the narrowest sense—another 150,000 or so people did die—mostly civilians, but it was mostly civilians in the whole war.

I have a complete rationalization for the thing, but despite that, my own feelings of concern and disquiet go back right to that day. I mean, I remember the day they dropped the bomb and I thought, we did it, and I was fully confident the war would be over. I wasn't following Potsdam, I didn't know about the emperor and the [Japanese] War Council— later information confirms my view—but I was fully confident that the

war was going to be over. But at the same time I did have a certain
amount of shivers and the feeling that this just isn't all good.

But he added, "We had to win quickly and decisively. Not *another* negotiated peace."

I knew York's well-reasoned arguments in support of the decision to
drop the bomb. I replied that although his analysis was persuasive, I still
had a great deal of respect for people like Wilson who seriously questioned its use. Although Wilson insisted to me that Bethe had more
effectively worked toward their shared conviction that Hiroshima never
be repeated, I, as an outsider, saw them as equally valuable pieces in the
difficult puzzle. Wilson's stance, I said, spoke very powerfully to the
"disquiet" York himself felt on learning of the atomic bomb's success.

I did not expect York's response. He told me that his arguments
were, in a sense, self-serving—that they involved self-rationalization.
"I'm very much aware of the fact that most sensitive, intelligent people
take more the Wilson view." When I asked York to explain what he
meant by "self-serving," he said,

> Well, I mean that I have to explain to myself and to you why I did these
> things. It's not only that I believe what I say is true. Even if I didn't
> believe it, I would be saying the same thing. In other words, like any per
> son, I have to explain myself to myself and to other people and this is a
> way of doing it. It's self-serving, not in the financial sense, but in the
> sense of providing an absolution and excuse.
>
> I have to tell you that when I wrote my book, *Making Weapons, Talk
> ing Peace*, I had a couple of goals in writing it. One of them was that I
> think that people are going to look back on this [the atomic bombings]
> as a terrible mistake and are going to say, "How could those people have
> done that?" So I wrote it, you know, with my grandchildren in mind.

York was actually puzzled about how he could have been thinking of his
grandchildren, because when he began writing the book he and his wife
had none.

> For years Sybil and I thought we weren't going to have any grandchil
> dren. We didn't know that for sure, but nothing was happening and we
> were quite satisfied with that. Now that it's all happened, we enjoy our

four grandchildren enormously, but back then we were satisfied and we never, never pushed the kids. The first one was born in '87, the book was published in '87. But I know that I had that in mind, but how could I? I can't quite put it together.

The other thing that I had in mind was I wanted it to be in the libraries so that future historical researchers and so on would have it as one of the things that they could turn to. I am acutely aware of the fact that there is something that needs to be explained. There's something to be explained because the natural point of view is to say, "My God, how could you do that?"

York has consciously and constantly chosen to remain on the inside of these troubling issues, convinced that only as a participant could he achieve any real understanding or influence. In recent years he has concluded that we can and should reduce the number of nuclear weapons on earth to zero. He has indeed been both a maker of weapons and a talker of peace. I came to York as an outsider with no technical or political expertise in nuclear matters. I had doubts about the use of the bomb and had long assumed that the only moral postwar response for scientists was to never again touch "the military atom." And I questioned the motivations of the scientists who have advocated arms control while remaining in the nuclear weapons–creating defense establishment. Thus our ongoing dialogue has been based on the mutual needs of successive generations—mine to understand and his to explain.

York had recently begun lecturing at the University of California about the decision to use the bomb. He told me that he thinks there is an unbridgeable generational gap between our understandings of the decision to use the bomb. He often opens his lectures by telling the students,

The first thing you knew about World War II is how it came out. And that's the last thing I knew about World War II. It took me four years to find out how it was going to come out. The first thing you knew about the atomic bomb is that we used it to kill a lot of people in Hiroshima. And that's the last thing I knew about the atomic bomb.

He is trying to convey the deep context in which the war and the ultimate use of the bomb developed. He does this first by describing what the world was like for President Truman, a member of the generation before York and thus twice or three times removed from his students.

> Harry Truman lived World War I, which I didn't, but I lived the Spanish Civil War. I watched in the newsreel Haile Selassie appealing for help when the Italians invaded Ethiopia just for the purpose of making empire, and nothing happened. I was aware even as a teenager, not of any of the details, but of both the prison camps in the Soviet Union—what we now call the Gulag—and the pre-Holocaust activities in Germany. And then comes World War II, which many people regarded as simply a failure in the peace of World War I.
>
> People love to say the Japanese knew they had lost. In fact, they only knew they had lost the remote empire. They had every good reason to believe that if they could let enough American blood, the Americans would negotiate. That was not a ridiculous or insane hope on their part.

York emphasized that he did not think the central question in judging the use of the bomb was whether or not the Japanese had already lost the war.

> Even at the beginning we never thought we'd lose. I don't know any Americans who thought, the Japanese are going to win. We always had enough self-confidence, based on reality, to know we could win. But, you know, it was gloomier and gloomier and gloomier for many years about how long it was going to take and how many dead there would be. And whether the Germans would win or not was even more of an open question, and whether all that would be left free in Europe was England. I think many thought, we can save England and not much else. And I've had this discussion to some extent with Barton Bernstein up at Stanford and on several different occasions and I find he's discussing the wrong numbers.

Bernstein, a Stanford University historian, received considerable publicity during the fiftieth anniversary of the atomic bombings for his arguments that actual military estimates for invasion casualties were well below the numbers claimed by Truman, Stimson, and Churchill after the war and accepted by most Americans as fact. In January 1995 the *New York Times* reported that his own calculation of Admiral William D. Leahy's estimates yielded 63,000 American invasion casualties. In an article on the same subject, Bernstein stated, "Given the patriotic calculus of the time, there was no hesitation about using A-bombs to kill many Japanese in order to save the 25,000–46,000 Americans who might otherwise have died in invasions. Put bluntly, Japanese life—including civilian life—was cheap, and some American leaders, like many rank-and-file citizens, may well have savored the prospect of punishing the Japanese with the A-bomb."[3]

I was interested to learn that York had spoken with Bernstein about this issue. He told me that although Bernstein was looking at Leahy's estimates for invasion casualties, these were not the numbers that mattered most at the time of the decision.

> There's fifty million people already dead, there's hundreds of thousands of prisoners, all of them Europeans because the Japanese didn't take Oriental prisoners, whose fate is uncertain, there's several hundred million people homeless, and you can't get at getting them back home until you get this whole thing over—*anything* that will win was "good," and if you say you've got an alternative, the answer is, we'll take them all, don't tell us you've got an alternative way, we'll take them both.
>
> And the other thing that some people treat almost as whimsical is the idea of an unconditional surrender. That was the most sensible policy you could imagine—especially for people of Truman's generation—not mine, but Truman's generation and Bethe's. World War II can be seen as a failure of the peace of World War I. World War II happened because we didn't have unconditional surrender in World War I. I think that's probably true, it's certainly a way of looking at it. It was no accident that von Hindenberg, who was the [World War I] German commander, is the senile president who appointed Hitler. It fits together.

When I asked York if he was able to convince Bernstein with his arguments, he responded,

> Well, no, but he will listen to me to a degree, and I listen to him to a degree. But he just focuses on what Admiral Leahy estimated the

casualties to be. Even if you take the Bernstein point of view on the estimates of what would happen in the absence of a bomb, what would it
really be like? It's not Admiral Leahy's memos or any of the others', it's
everybody's vivid memory of Okinawa. And the fact the Japanese never
surrender, and they put civilians in the line of danger, deliberately, that's
what was in everybody's mind. Now there were numbers attached to
that, Okinawa being such a small place, but even there the number
killed is whatever, 40,000 or 50,000. Not Americans, but Japanese,
and in such a tiny place. And the Japanese never surrendered.*

Although we discussed the disagreements among historians about the
invasion casualty estimates, York's primary argument against this numbers game is even more basic. "Well, I claim they aren't even the appropriate numbers to consider; fifty million is the number to focus on."

As we spoke, it occurred to me that there has always been a discrepancy between what military planners have hoped for, and therefore predicted, and what actually has happened in battle. Contrary to confirming that such estimates are realistic, history would seem to teach just
the opposite. How many times had monarchs or generals planned to
win battles or even wars in months, only to have them go on for years?
The slaughter of trench warfare in World War I is a prime example.
Richard Kohn, the former chief of air force history, wrote of the controversy over invasion casualties,

> I've always thought the casualty argument . . . simple-minded and
> lacking in context. The real issue was the campaign, not just the
> invasion. With Japan prepared to fight all-out indefinitely, and hav-
> ing stockpiled 9,000 aircraft, most for kamikaze use, the casualties
> would have been just tremendous on *both* sides, and everybody at
> the time knew it. That should be explained; planners put in num-
> bers because that is necessary for logistical and other reasons, but to
> argue about them is utterly to miss the point, and that is what
> scholars have done. The controversy over numbers trivializes the
> business.[4]

---

* American combat casualties on Okinawa exceeded 49,000, with more than 12,500
dead. Japanese combat casualties are estimated at 90,000. Okinawa was defended by
77,000 Japanese army troops and 20,000 Okinawa militia, and even Okinawan children.
Only 7,400 survived to become prisoners of war.

Outside of the question of invasion casualties, the assertion that the bomb brought the war to a swift end often led me back to the question of why it had been developed in the first place, and how its meaning changed over the course of the war. I struggled with the shift away from the Allies' original concept of the bomb as a necessary *defensive* weapon to counter a Nazi atomic bomb. In 1939 scientists and statesmen in the know were frightened because they understood that the Nazis were evil enough to use such a terrible weapon. Yet by 1945 the bomb became, for the Allies, an *offensive* weapon whose ultimate rationale was that it would end the war quickly and decisively.

I wondered what York, a longtime Pugwash participant, thought of Joseph Rotblat's views of the atomic bombs and the scientists who built them. He responded that although he considered Rotblat a hero in general, he strongly disagreed with him on this point.

It isn't that they [the scientists] shirked the responsibility, they wouldn't have admitted there was such a responsibility. So, I realize—he [Rotblat] is unique in that regard. And it *wasn't* a shift. What makes it look like a shift is that we weren't in the war in the beginning. That is, the early decisions were made at a time when we were not in the war with the Japanese. So we were not thinking generally or clearly about us defeating somebody. We were only thinking about maintaining our safety against the possibility of someone else getting the bomb first.

The change in thinking about the bomb did not happen gradually. It happened on December 7th [1941] when we got caught up in the war. Then everybody turned around no matter what they were doing, not just those working on the bomb, but *everybody*, and said, now this is our war and we've got to win. And then the fact that it went so poorly for the next couple of years. It wasn't that most of the navy was sunk at Pearl Harbor; the Japanese also took all of these islands, Guam, the whole Dutch East Indies, as we called them, Hong Kong, Singapore, Malaysia, they were in New Guinea poised to go after Australia, or so it looked. And it really looked as though we had a terrible, tough time ahead of us. Well, that was correct, we did. But the change in thinking about the bomb didn't happen gradually.

Now, I'm oversimplifying, but I don't think that I'm oversimplifying much to say that before December 7th, if you'd asked somebody why we're working on an atomic bomb they would have said, well, we had better make sure we do this before the Germans do because that's where fission was discovered, and so forth. But on December 7th people would have said, we're working on this bomb in order to win this terrible war. But it wasn't a gradual shift that happened when nobody was looking. I mean we were all looking very clearly on December 7th.

I don't know who would have used words like this, but the idea was to crush them [the Japanese] because World War II had grown out of World War I. I think there was a very wide awareness of that, so it wasn't just, win World War II, it was, win World War II in such a way so that there won't be a World War III. And that's a sufficient reason for doing anything. And the idea that now that the Germans are defeated, there's no more purpose, that has never made sense to me. I didn't know about anybody saying it, but it wouldn't have made sense to me then, and it doesn't now.

However, a little later in the conversation York raised another point, perhaps in response to Rotblat's claim that science should never be used for destructive purposes.

There is something else here that relates to this question of something that isn't guilt, but of concern, some other word. The idea that we had to do this before the Germans did was another argument in favor of doing something which obviously was outside of what science normally does. So I'm not sure, but I have this feeling that even in my own case, I may have used that to explain to myself why we had to build this huge bomb.

I told York that we often seem to focus on the responsibility of the scientists for having created the wartime weapon while the soldiers, their fellow citizens, are expected to kill and are honored for doing so. Some find it deeply disturbing that scientists used their knowledge to build the bomb; others criticize any scientist who expresses concern or regret. After all, they assert, during war people kill and are killed. Why should the scientists hold themselves above and exempt from such worldly realities? York said he had a theoretical concern about this problem that was along an entirely different line. I was eager to hear it.

As modern civilization has progressed it has become more and more difficult to kill somebody face-to-face. In order to get Americans to kill other people you have to go through an extensive training with marines sticking bayonets into boxing bags and things like that. It takes a lot of work to persuade Western people to kill other people—even in a war. And most of them don't do it, I don't know what the data is. But there's a lot of reluctance.

But the further you can back off, the easier. So bombing is easier than bayoneting. Even the use of cannons and artillery is easier. It's bayoneting that's hard, and it gets progressively easier as you make it more remote. The atomic bomb simply makes it much easier still. I don't know how important that argument is, but it is something I have thought about, that one of the special evils of the atomic bomb is that it just makes it easier. It isn't the matter of the Ten Commandments; they have been there for three thousand years, but life was tough and life was raw and as life gets to be better and everybody has it easier, it's getting harder to kill.

But we have invented these mechanical ways of which the bomb is simply the extreme, in which you are far enough away so that you don't have to look them in the eye. To bayonet somebody you have to look them in the eye. Now what makes bayoneting possible is when the other guy is about to do it to you, which can happen as well. But in Nanjing [Nanking] and places like that, the Japanese just killed women and children with bayonets without any threat at all. Maybe from their superiors, but not much of a threat.

I find it easier to forgive the Japanese than the Germans, because the Japanese were still coming out of the Middle Ages and the common code, I mean, the "good people," the mothers, the children, the philosophers, all had these ideas about the glory of war and killing for the emperor as being a high and noble deed and military death being a glorious death and so on. That was part of their total philosophy. Whereas the Germans were a hundred years past that and they did it anyway. They were among the world's most civilized people, and the German philosophy included all sorts of modern ideas. Also the German behavior is something that we should all be more concerned about because we are like them.

York applied his analysis not only to the past but to the present and the future as well.

There are several problems that we are not coping with. One of them is that the more mechanical we make it, the easier it is to do, because you don't have to look them in the eye. And besides that, Americans have been the most successful fighters in the world at killing—a high ratio of killing other people. We always have the fewest casualties in every war we get into, by a large margin. And both of those things make it easier for us to make war than I'm comfortable with. We can get away with it, our people don't get killed in these wars.

York himself was a player in the scientific, military, and industrial system that invented easier and easier ways for us to kill people, and he has been crucial in the efforts to control that capability. At the end of World War II, he returned to Berkeley. There he earned his doctorate, conducted research, and began teaching. In 1949 the Soviets exploded their first atomic bomb, and in January 1950 Truman announced that work on the hydrogen bomb would go forward. Early in 1950, at the suggestion of Lawrence and Luis Alvarez, York and a Berkeley colleague went to Los Alamos to explore the matter of the H-bomb. At that time York knew little of the secret debate and controversy regarding the thermonuclear weapon's development. Seven years after the end of World War II, Lawrence handpicked the thirty-year-old York to serve as the first director of the new weapons lab in Livermore, California. Among those who came to Livermore was the mercurial and brilliant Edward Teller. Convinced that Los Alamos would not carry out the crucial Superbomb work properly, Teller had lobbied hard and won his bid for a second weapons lab. "When we started the Livermore laboratory," York laughingly told me, "he was the oldest one around. We thought forty-four was pretty old." When I asked him about his postwar weapons work, York explained,

I don't think I've written this anywhere, but I've implied it. Part of my willingness and even eagerness to set up Livermore was that I wanted to

Figure 17. Ernest O. Lawrence, Edward Teller, and Herbert F. York, Livermore, 1957.

know what was going on too. Here was a whole branch of science that was sort of hidden from me. I learned about it not so much at Livermore as when I went down to Los Alamos the year before.

Did he want to learn more about the politics or the science of weapons?

No, no, I wanted to know the *science* and the *scientists*. You see, I knew that Fermi was there, Bethe was there, von Neumann was there, [physicist John] Wheeler, Teller was there. I hadn't really had a good chance to know those people, so I jumped at the opportunity to go—to find out how these bombs really worked, how it really was at Los Alamos. At Oak Ridge I really hadn't known many details, and then all these great men were there. And it was just a joy to contemplate getting to meet them. So there is that attractive and seductive side. But the other piece to it, I don't know whether Bethe would say this, but it was Fermi's view, to the extent he had political views, that in a democratic country when the people, when the process decides that this is what you are going to do, then you have a certain obligation to participate. And there's a certain humility there.

I generally agreed with York on this but added that the dilemma always remains whether to stand in opposition to the will of the people for a deeply held conviction. He replied,

> There they make a distinction of the fact it's a democratic government. They could all question why Heisenberg helped Hitler without ever applying that question to themselves, because they all believed, ours is a constitutional democracy and there are ways to decide these things and I have my voice but that's all I have.

Later, when I was alone thinking about these questions, it occurred to me that such logic does not necessarily hold. An essential characteristic of the nuclear weapons world is that so few people have access to classified information. Most of us are on the outside. Fermi, who had opposed the H-bomb development on moral grounds, worked on it once Truman decided to go forward with the program. Although he may have believed that he acted conscientiously, the process, cloaked in secrecy as it was, was far from democratic.

York subsequently brought up this same point during our general discussion of the relationship between specialized scientific knowledge and society. It was just after the news of the sheep cloning had broken. My concern was that the scientists would remain isolated in their own exciting world, proceeding as they pleased, without adequate or informed public debate about what should be done. I feared that we would ultimately be left with whatever the scientists chose to make from their knowledge. York replied,

> Your study is related to that because you have the fact that it really was just Roosevelt, Stimson, and Marshall who were deciding everything at first and then, later, Truman. And as for deciding on whether there should be a [nuclear bomb] program or not, you have this one S-1 Committee of Compton, Lawrence, and Oppenheimer, plus a few others like Fermi and Szilard. I mean, kind of a floating craps game, not a particular, easily identified group, but a group with fuzzy edges, with a small group of people at the center who are deciding. And the same thing happened again with the hydrogen bomb. A relatively small group

of people, bigger than the first group, but essentially just a handful of people deciding all of this.

Then there are those people who insist that progress is inevitable. It's a phrase one keeps hearing. I think that when progress depends on one person or three persons talking and thinking among themselves, it probably is inevitable, in some sense. But when it takes a billion dollars of government money to make something happen, then it's not inevitable.

On several occasions I told York of my discomfort with scientists as defense department insiders. By participating, hadn't they essentially contributed to the problem of the arms buildup? He responded on one such occasion,

Did I tell you about my conversation with Rabi? It was some ten years after the Oppenheimer case, maybe more, and we were talking about the Oppenheimer case. Rabi was lamenting the fact and he said, "You know, I replaced Robert as chairman of the General Advisory Committee; I don't know why they did that because I had the same views that Robert did, but even so they made me chairman." Of course, he accepted being chairman. And then he added, "You know, if I had made a big fuss about it with highly critical statements about the AEC [Atomic Energy Commission] and so forth, if I had quit in a huff I would have got a couple of lines in the *New York Times* and that would have ended my ability to have any influence at all over what was happening."

During my interviews with the scientists, I often pondered how being the agents of such profound historical change affected them. They were caught up in the forces of history but at the same time had made choices that changed its course. David Hawkins asserted that the atomic scientists' lives were irreversibly altered. Some see them as burdened by guilt; others, including some of the scientists, scoff at such notions. Yet they lived at an intersection of historical coincidences over which they had no control. During our conversations, York considered the question of responsibility within the larger context of historical coincidences, two of which he explained to me. I had thought about the first one, and had discussed it with other scientists—that fission was discovered on the eve of World War II. A second coincidence was new to me: the relationship of the making of the bomb to the age of the Earth. York asked me to reflect on both.

> Try to imagine what it would have been like if they had discovered fission five years earlier; it could easily have been so. Or ten years earlier, where you would have to change a lot of other things as well. Or five or ten years later. It just wouldn't have had the same kind of relationship to World War II.
>
> Then the other coincidence is the grander one. If the Earth was a billion years older, you wouldn't have been able to do the Manhattan Project. You'd still have been able to do all that technology, but it would have been very much harder and we wouldn't have been able to do it during World War II. The reason is that the half-life of U-235 is much shorter than the half-life of U-238. Ordinary uranium in the ores everywhere in the world is .7 percent U-235. That ratio changes with time. And roughly speaking, it changes by a factor of 2 every billion years. The Earth is 4 1/2 billion years old, the material out of which the Earth is made is something older than that.
>
> If the Earth had been a billion years older, we would still be able to do everything that we are doing today with atomic energy, atomic weapons, and so on. But it would have been much harder to get it started, because the reactors of the type that they built at Hanford would not have worked if the Earth had been a billion years older. And the separation of uranium into its isotopes would also have been much more difficult if the original stuff was only half as rich. So the Manhattan Project would have been much more difficult to do and very probably couldn't have been accomplished during World War II.

Returning to the coincidence of the discovery of fission on the eve of World War II, York reminded me that Fermi had been working on the problem in Rome in 1934 and 1935. I said it was my understanding that Fermi had probably achieved fission without realizing it. York then remarked that the obvious is never as obvious as we would like to think.

> People would look at that and say how it's obvious. He should have, but he didn't. And neither did the rest of the group. And then the rest of

the group got broken up before they could finish. A lot of things that ought to be obvious really aren't, even to very good people.

The development of the hydrogen bomb, York told me, was an example of the same problem.

Well, I looked at the *George* device and asked myself, why didn't the group of people [at Los Alamos], why didn't even myself, looking at that, invent *Mike* instead of just Teller? And even Teller had been staring at these ideas for years without making just a very small extension. Why didn't [physicist Richard] Garwin? Why didn't Fermi? When I went to Los Alamos in the period 1950–51 this great team was there. Fermi was there and Garwin was with him and [physicist] Leona Marshall too, and then Bethe was there a lot of the time. I'm not sure who was there part time or who was there full time, but John Wheeler was in and out, Johnny von Neumann was in and out, George Gamow was there as a frequent visitor, and they're all of them trying to figure out how to make a hydrogen bomb, not succeeding. And all you have to do is start from the *George* experiment and make some minor changes and there it is. Why didn't any of them do it?

Exploded in May 1951, *George*, while not a prototype for an actual hydrogen bomb, was designed so that the scientists and military could learn as much as possible about thermonuclear reactions. Eighteen months later, *Mike*, the first large thermonuclear device, was tested, yielding more than ten megatons, one thousand times that of the Hiroshima bomb.[5]

York told me of the night that the secret of the H-bomb was revealed to him. It was a warm tropical evening on Parry Island. The island is part of the Eniwetok Atoll, which lies in the Pacific, about twelve hundred miles south-southeast of Tinian, from where the atomic bombing raids were launched on Hiroshima and Nagasaki. It was just five years after the war, and the reef surrounding the atoll was still scattered with the remains of ships lost to the vicious Pacific battles. In a corroding aluminum building, Edward Teller stood before a blackboard and sketched for his eager young colleague the key to the Superbomb. York saw immediately that Teller had solved the problem and simultaneously felt something that he has difficulty explaining. "I don't think it's guilt," he said, "but it's something. It's that I realized it was too bad." When I asked York what had disturbed him at that moment, he answered,

> It was the ominous nature of the thing. Let me go way back. There was a visitor just here, Rich Wagner, identified as being very hawkish. He's active as a consultant on nuclear things in Washington, and he was formerly at Livermore. And we had a meeting here talking with a group of people—about history. And I have often said that if by magic you could eliminate the process of fission, that I'd favor doing that. That the world, I mean the human race, would be better off without fission. Even though I think it has a lot of benefits, the net benefit is negative. But I've been saying that for quite a long time as a way of describing how I feel about it.
>
> But Rich Wagner posed it to me as a fresh question. He put it a little differently. "If you could change the cross-sections so that you couldn't build the bomb, would you do it?" I said, "Yes, I've thought about that before." He said, "So would I." I thought of him as hawkish, but he also feels, felt or said, that you know if you could get rid of the whole thing you'd be better off.
>
> But it's not guilt, it's just disappointment with the way nature was designed and a feeling that it was designed in a way that's ominous, or at least it contains these ominous features. So I don't think I have any feeling of guilt. But, of course, you can't—you really can't probably analyze yourself accurately.

Earlier York had alluded to Hans Bethe's statement "When I started participating in the thermonuclear work in Summer 1950, I was hoping to prove that thermonuclear weapons could not be made. . . . I never could understand how anyone could feel any enthusiasm for going ahead."[6] York told me that he found it "somewhat naive" for a person as smart as Bethe to reason along those lines. "Proving that something like that can't be done is usually not possible when all the fundamentals are there," he asserted. Perhaps, I thought to myself, it was not a matter of intelligence: Hans Bethe had hoped the Superbomb could not be made; Herbert York felt deep misgivings when he realized it could.

It is interesting to note that two men who played key roles in the British A-bomb project had reactions similar to York's on first learning of the possibility of nuclear bombs. Sir Henry Tizard, who was to head Britain's early research into atomic weapons, asked, "Do you really think the universe was made in this way?" And for the British physicist Frederick Lindemann (Lord Cherwell), who became Churchill's personal science adviser, "the idea of such destructive power being available in human hands, seemed to repel him so much that he could scarcely believe that the universe was constructed this way."[7]

To York, nature seems ominous. It occurred to me that as long as we settle our differences with violence, we are bound to use whatever nature provides to win the advantage and defeat our enemy. Just as Prometheus stole fire for humankind, science stole the power of the atom, and humans would use it as they pleased. Was this the reason York believed that we needed academic scientists on the inside? Was it because if we were working toward the end of reducing nuclear weapons, good technical people were needed to solve the problems?

> Well, I claim the goal is more general. General war is the problem, and within general war, nuclear weapons are only another factor. The whole idea of attacking populations—that's what the Mongols did in the twelfth century—seemed to have disappeared with knighthood. But we are now back to where you kill whole populations, and the atomic bombs didn't start that. So the object is to avoid World War III.
>
> Whenever I hear the Japanese saying that there must be no more Hiroshimas and the way you make sure there are no more Hiroshimas is to explain how bad Hiroshima was, I say that's nonsense. If you don't want another Hiroshima, don't make another great war. If there's another great war, there will be more Hiroshimas because people aren't going to sit back and remember how bad Hiroshima '45 was, they're going to focus on the terrible mess they are facing right then and they are going to turn to anything to end it. So if you don't want another Hiroshima, don't make another great war. That's the whole solution to no more Hiroshimas. I mean that's my view.

I said that one of the strongest arguments for remembering Hiroshima is that it helps to remind us of how terrible war is. After all, the world jumped from World War I, with its horrendous casualties, directly into World War II. York agreed.

> That helps to remind you of the whole war. But I would just as soon show a picture of—well, there are no pictures of Nanjing, or the whole fire bombing of Tokyo, there are pictures of Dresden and Coventry. The Germans deliberately destroyed the center of Rotterdam as an act of terror to persuade the Dutch not to fight, and it worked.
>
> I think people should remember. But you know the movies of Berlin with these women and children picking through the rubble trying to find food are at least as moving as Hiroshima is. I think it makes sense for Americans to express a certain concern and regret and a "never again" spirit with respect to Hiroshima, but I find it offensive when the Japanese do it.

Since this meeting with York, two volumes have been published that contain photographs of Nanking. What they lack in number, they make up for in horror. Iris Chang cited estimates of noncombatant dead in Nanking ranging from 260,000 to more than 350,000 in a matter of weeks. She also asserted that the number of women raped and sexually mutilated was one of the greatest in history. John Rabe, a German in Nanking at the time, recorded in his journal that on a single night, 1,000 women and girls were reported to have been raped.[8]

Chang drew parallels between Nanking and the slaughters of Hitler and Stalin, which took years to accomplish, and graphically described the cruelty of the killings. And she pointed out that unlike the bomb or the Holocaust, the atrocities in China are little known outside of Asia. She argued that the Japanese have cultivated the postwar view that they were the victims, not the instigators, of World War II and that "the horror visited on the Japanese people during the atomic bombings of Hiroshima and Nagasaki helped this myth replace history."[9] Chang's work sparked a heated debate in the United States and in Japan about the facts of the massacre and the actual casualty numbers.

Asada Sadao, who examined the differing Japanese and American views of the bombings, observed, "Now that the Cold War is over, Cold

War revisionism seems to have lost much of its relevance. The Japanese must also guard against what has been termed 'A-bomb nationalism,' which claims for Japan a morally unique position as the only victim nation, for such an attitude can only militate against the universal appeal for nuclear disarmament."[10]

I wondered if York was offended by the Japanese attitude of "never again" toward Hiroshima because of the atrocities they had committed.

> It's not just what they did, it's the fact that they are on the wrong track. If they want to avoid another Hiroshima the way to do it is, don't make fifteen years of war on China, you know, and on the rest of the world. That's the way to avoid another Hiroshima.
>
> One of the distinctions among intellectuals that you will see if you present any of this discussion to somebody from the philosophy department, the literature department, sociology department, they will say, "Well, if it's wrong to work on those things, you shouldn't and that's it." Whereas the people I know, and I think nearly all physicists would be like this, but certainly the ones that we've been talking about, Bethe, Rabi, and all the others, say that in order to have any influence over what's happening at all I have to be part of it.
>
> I believe that the resolution of all of these problems facing us takes time—I thoroughly believe that. We've been working on becoming civilized 2,000 to 4,000 years and we still have a long way to go. And that the best way to make a contribution is to be in the mainstream. The action's in the mainstream. That's where most people are, including most good people, most smart people, most passionate people, that's where most everybody is, if you want to make a contribution, you've got to be there. Then once you're there, when you are somebody like Bethe or [Stanford physicist Sidney] Drell or Garwin, whom people want to consult because they have credibility in certain areas, you should do that. So I was very sympathetic with Rabi. I would have done the same thing.

York has been an insider since he joined Ernest Lawrence's circle a half century ago. In 1960 he was instrumental in forming JASON, a group of first-rate scientific researchers consulting on defense and national security problems for the government, using the most up-to-date science. JASON consults with various departments and agencies of the government, including the Departments of Defense, Energy, and the Navy. The group did important early work on arms control in the 1960s but became notorious during the Vietnam War for an "electronic battlefield" project that would have installed special sensors in the jungles of Vietnam, thus denying the enemy the protection of natural cover. Since many JASON members were college professors, they were subject to intense criticism, particularly on campuses, when their roles became public.

Several leading scientists left government service during the Vietnam War. Among them was George Kistiakowsky, the outstanding Manhattan Project member and much-esteemed former presidential science adviser. He refused to participate in the defense establishment for the rest of his life. I recall my parents, who were strong opponents of the war, discussing Kistiakowsky's decision and telling us of the high regard in which they held him.

For his part, Herbert York felt pressure to resign from JASON, both from his friends and from his conscience. In *Making Weapons, Talking Peace*, he wrote,

> I believed that the [Vietnam] war was a bad mistake, that our cause was hopeless, and that by continuing to fight we would only prolong the misery and increase the death and destruction. . . . [I did not] think it was a moral issue except in the general sense that all wars, especially modern ones, involve important moral questions and deep ethical contradictions. . . .
>
> Despite my total lack of empathy with the main action going on at the moment, I continued to believe that the defense of the United States—and thus of the West—was a most worthy goal, and I determined to continue all my remaining relationships with the defense establishment, including *Jason*. . . .
>
> Many loyal and patriotic people did otherwise. . . . In so behaving, they joined another group of veterans of defense science and technology—including Philip Morrison, Victor Weisskopf, Robert Wilson, and many other Manhattan Project physicists who much

earlier, in the very first postwar years, decided they had done enough, or more than enough, of that kind of work.[11]

While committed to a strong national defense, early on York was also deeply concerned that the country's ever-increasing military power was accompanied by ever-decreasing national security. A long-time arms controller, now that the cold war is over, he stresses that we must push hard to drastically reduce the number of nuclear weapons. Unlike some of his colleagues, who insist that the maintenance of deterrence is imperative, York's immediate goal is one hundred nuclear weapons, and ultimately zero. He is concerned that although we are reducing the numbers, the focus of the U.S. weapons labs is too much on maintaining a safe stockpile and not enough on substantially decreasing it. When I asked what he thought of the current program to maintain the safety of the nuclear arsenal, he said,

> As part of the American political system it just absolutely requires it. I claim it's not necessary in some more abstract level. But it is necessary given the world the way it is. The only way I can fit the whole picture together is to say that at the same time as you are doing that, you need to be really pushing hard on the minimization of nuclear weapons, of reducing the numbers, reducing their role. The Council for a Livable World produced a list of things that should be done and one of them is that training and planning should exclude them [nuclear weapons], because if you train the troops in their use it is just one more situation that makes it likely to use them.[12]
>
> Well, you have to proceed in a way that makes it seem realistic to the rogue states that we would use them. You have two situations, maybe more, but two important ones. One is the case of the Iraqs and the North Koreas. Another situation is that any leading, any big industrialized country such as Japan—let's say Japan and Germany—could produce nuclear weapons from a standing start in just a few months.
>
> But I do agree that the goal should be zero. And that you can't—there is no way to achieve that goal in the present world. You see we are currently talking about 10,000 nuclear weapons on each side. We pretend it's only around 3,000, but in fact it's around 10,000. But everybody is self-deceived—either deceived or self-deceived about that.

It's tricky because there's "strategic" weapons, which is what people focus on, then there's so-called tactical, there may be some other name developed for these—"theater" or something—and then there's "active reserve" and "inactive reserve." And for now, it's even worse than that because neither the Russians nor ourselves have destroyed many. There were 35,000 here and there were 45,000 there. And we don't even know what the real totals are there, especially in their case [the former Soviet Union], but you don't really know what they are here either. The process of getting rid of them is so complicated that it would take a team of really good people going around everywhere to discover all of dozens of intermediate states on the way from full deployment to destruction and somehow summing over all of them.

One of the many things I learned from talking to and reading the works of insiders, even if they are not allowed to tell me any of their big secrets, was that they could certainly reveal the extravagantly complex web of potential death making that has developed since the *Enola Gay* dropped her primitive A-bomb on Hiroshima. Of course, many would argue that it is not death but deterrence that these weapons have actually created. Nonetheless, the magnitude of the numbers and the institutions that have grown up around nuclear weapons are overwhelming to contemplate—from the outside it looks like manifest insanity. An example: Richard Garwin sent me an article he wrote with Sidney Drell about basing the MX missile on small submarines, and another on reducing dependence on nuclear weapons. Both of these pieces gave me a concrete sense of the depth, complexity, and entrenched nature of nuclear weapons in our national defense. As I earnestly read these documents, it seemed to me that the numbers and names of all the nuclear hardware would have been laughable if they had not been so real. In the margin of one article, I wrote "ignorance is bliss." With the realization of the absurdity of the numbers, I wanted someone to blame. You guys made this mess, now you fix it! I could turn York's phrase around and say, "The last thing you knew was that you had created a world that can be destroyed by nuclear weapons, and that's the first thing I knew." Yet, in 1968, Garwin and Bethe wrote an important analysis and critique of antiballistic missile systems, exposing how easily penetrable they actually were.[13]

These scientists have chosen to do more than their physics. They have become involved in national affairs from a sense of responsibility. In 1994 a JASON committee studied the need for nuclear testing. Drell sent me a copy of the committee's summary and conclusions, which, after making a logical, technical argument, showed that the United States could enter into a comprehensive test ban treaty of unending duration.[14] It was clear to me as I read the document that only those with technical know-how, those proven and trusted by the government, could have made the logical argument that would lead to this conclusion and remain credible. Only conscionable "insiders" could have done this.* Yet insiders are often deeply distrustful of the ordinary citizens of our democracy. We are kept outside through the governmental institution of secrecy.

* The efforts of such insiders notwithstanding, in October 1999 the U.S. Senate voted against the ratification of the Comprehensive Test Ban Treaty. It was the United States' first rejection of a security-related international treaty since the Treaty of Versailles, eighty years earlier.

✴ Outsider History

One evening, resting on the floor after a day of reviewing the tapes of my conversations with Herbert York, I began to question my own attitude toward the "insider/outsider" question. Why was it, I wondered, that I had this almost blind reaction against scientists working in defense? I thought of York and his scientific colleagues who, in addition to doing their research, have dedicated their lives to using their technical expertise for what they deeply believe is the good of the nation. I could either agree or disagree with their views and appreciate how their thinking had evolved over the years.

I have always tried to be a good citizen. Although involved in my community, I have never done anything as important as the scientists' work. I asked myself, who am I to criticize? I respected their dedication—knowing full well that the power, prestige, and influence they wielded as members of an elite group were no small matter. But why, on a deeper level, was I suspicious and distrustful? This reaction did not match my actual experience of a man like Herbert York, who clearly has labored long and hard on difficult and complex issues and whose aims I share.

As I thought about this, I experienced what I will describe as a "flatness" in my brain. It was as if there was a dead spot in my thinking process on this question of being "inside." As I focused on this feeling, something unexpected happened. I remembered my high school friend Bruce Richardson. While I always say a prayer for him on Veterans Day, I had not really thought about him for years. My eyes slowly filled with tears. A few overflowed and ran down my cheek, trickled behind my ear, and got lost in my hair. Bruce was killed in the Vietnam War.

We were not close friends. He was just a sweet, simple boy from my hometown. We did not naturally fall into the same group. He was poor and not a good student. Although I was not wealthy, I was the daughter of a Brookhaven lab scientist and lived in a nice, big house across from the high school. But neither Bruce nor I ran very much in the crowds we had been assigned to. Ours was a small town; we had been in school together for years and, being nice kids, were always friendly with each other. And in 1967 many of us crossed established boundaries on bridges built of music, adventure, and fear.

My mother codirected and choreographed our high school plays, and in my senior year we did *Oklahoma!* She cast the leading roles from the school's usual musicians and scholars but needed boy dancers to play cowboys and ranch hands. One afternoon she called an audition, and naturally, lots of boys came to watch the girls, including me, dance. One group lounged in the back of the auditorium with no intention of participating. But my mom, drawing herself up to her full five feet, looked to the back of the hall and called to the boys in the voice no one dared ignore. Soon we were laughing and whirling around to the strains of Richard Rodgers. Bruce was one of the boys my mom caught, and so we became better friends.

Then, one summer night, after graduation, we met at a party and danced. I had a crush on another boy and waited all night for him to show, which he never did. Bruce knew about the crush but stayed with me anyway. We went outside and he tried to kiss me. I did not want to—it was unexpected and I did not know what to do. I remember his tall, skinny body standing close in the dark and his voice whispering, "Oh, come on, Mary." But I was shy and still said no. So he walked me home.

Bruce did not graduate. He had failed too many classes and was supposed to complete them the following year. In the fall I left for college in California, and we never saw each other again. But one night I dreamed of him: I was with my best high school friend, Christine, looking at clothes in a department store. Bruce came in and said he was glad to see us. We greeted him and talked just as we had done so often back home. He told us he had come to say good-bye, turned, and walked through the wall.

Soon after the dream, my mom called to tell me Bruce had been killed in Vietnam. She said he never completed high school, had been classified 1A, and lost his life just a few weeks after arriving in that unfortunate land. I did not cry then. I do not know whether he became a killer before dying. It took me nearly three decades to realize that when I first learned of his death, some thin veil had descended, some small part of my heart had turned away.

I wondered how it could be. He was one boy. Many have lost much more than I. So many Vietnamese were killed, and the fighting was on their homeland. I was raised in a safe environment. I have never suffered as my parents' and grandparents' generations did. I am just too soft—that's life and that's war. It was, after all, a small war by comparison. Bruce was not my brother or my father. He was just a hometown boy I never kissed.

But so it was with me. Lying on the floor, I experienced the confused grief of my long-ago self, and finally cried for that boy. Then I realized something else. In war, don't people usually blame the enemy for the death of those they love? When my mother told me that he was dead, it never occurred to me to blame some North Vietnamese soldier or the Vietcong. I blamed my government for wasting Bruce's life, and in my own young way, right or wrong, I took another step to the outside.

✳ Running to Ground Zero

Thirty years later, in October 1998, I traveled to New Mexico to see the atomic bomb test site, located on what is now the White Sands Missile Range. The area is closed to the public, but twice a year the army allows visitors into Trinity. I booked a room in Socorro, about thirty miles from White Sands's northern entrance. The day before the site's "open house," I flew to Albuquerque, rented a car, and headed south on Interstate 25. There was little traffic, and I cruised easily past Isleta Pueblo and Belen, arriving in Socorro in a little over an hour. The marquis in front of the Holiday Inn announced "Trinity Site Open Oct. 3." In the lobby, I registered and picked up the Socorro paper. On the front page, below the news about the blessing of the animals at a local church, was an announcement of the test site opening. I verified the directions, explored the town, and ate an early dinner. Before going to bed, I watched the moon rise over Socorro from the landing outside my room.

I was awake before dawn the next morning. Trinity would be open between 8:00 and 2:00, but I had heard that the line at the gate could be long, and judging by the previous day, it was going to be hot. Maps and directions in hand, mileage calculations noted, I set off in the still, cold darkness. I headed south on the interstate and turned east at the little town of San Antonio. As I drove down the lonely state road, the sky began to brighten, but the distant low mountains blocked the sunrise. Then, rounding a bend, I faced the biggest sun I have ever seen, resting low and heavy on the horizon. Blinded by its brilliance, I instantly thought of the bomb, and how much brighter it must have been. Next I remembered the name of David Hawkins's home village near Alamogordo—Nuestra Señora de la Luz, Our Lady of the Light.

I slowed down and peeked at the road, tears streaming down my cheeks. With my lids half shut, my right hand and my baseball cap shading my burning eyes, I carefully navigated until I reached the missile range sign. Then I turned right and drove south toward the Stallion Gate.

Soon I came upon a handful of cars parked off the road near a chain-link fence. Several people were outside, chatting and making introductions. I joined them, each of us anticipating our own, personal Trinity. Ahead of me in line I noticed a couple I had seen in the motel lobby the night before. The woman had Asian features, the man European. I had over-heard them talking and, judging from their accents, figured that her native tongue was Japanese, his American English. I assumed they were married. The woman was now standing alone, wrapped in a shawl, gaz-ing east toward the gray mountains and the rising sun.

Cars continued to line up behind me, and at 7:45 missile range staff, both military and civilian, gave us our instructions. They handed out the entry rules, an information booklet, and a sheet explaining radia-tion with a chart for calculating radiation exposure, included to demon-strate that most of us have an exaggerated fear. We also received forms releasing the U.S. government from any and all possible claims. We were instructed to sign and return them at the entry gate, where the guard counted to make sure that she had a release for every woman, man, and child in each car. The bomb test site was seventeen miles from the gate, and the crowd of cars spread thin along the desolate, yucca-lined route. We were allowed to drive only to the parking area near ground zero; all other roads were blocked. At the entrance to the lot, soldiers directed us to the exact space in which to park. Mine was about the tenth car in. After getting my bearings, I passed through a gate and felt a dull sadness. It was a quarter-mile walk to ground zero.

Ahead of me was the woman who had been wearing the shawl at the Stallion Gate. It was warmer now, and she was in a T-shirt, jeans, and hiking boots. Her jet black hair hung straight down her back. Her hus-band was beside her, but as I watched, she began to walk more quickly, and soon pulled ahead of him. The path began to fill with people, and she seemed to be trying to stay in front of the crowd. Her brisk, uneven steps became a deliberate jog, and she left him behind. The woman was running to ground zero.

I reached the entrance and paused to read the sign warning us not to eat, drink, chew gum, or apply cosmetics. We were also instructed not to touch any Trinitite, the desert sand that had been transformed by the bomb's heat into smooth, jade green stones. In the early 1950s the Atomic Energy Commission, in a cleanup effort, scraped away and buried most of the Trinitite.

Seeing the sign, I remembered that my mother, like so many project scientists, had collected these glassy talismans of Trinity when she was given permission to visit the test site a few weeks after the atomic blast. I recalled a story her younger sister, Joan, told me. My mother brought the Trinitite home to Chicago in a cotton-lined film reel can and placed it in an antique cabinet in her mother's living room. She instructed her family not to handle the rocks, but from time to time, when no one was looking, Joan, who was sixteen at the time, would quietly open the container, remove the top layer of cotton, and finger the Trinitite. Aware of the danger but fascinated by the stones' deep-glazed hues and translucent beauty, she would carefully turn them over in her hands. Now the Trinitite lies crumbling in my younger brother's desk drawer.

When I stepped through the gate at ground zero, I saw the large circular fence surrounding the area and thought, Stonehenge. I noticed only the slightest depression in the earth as I walked toward a rough-hewn, volcanic rock and cement obelisk. In the distance, across the parched wilderness beyond the fence, rose the Sierra Oscura range. People began to mill around and snap pictures. The woman stood in front of the obelisk, reading the inscription. Her husband was on her right, and I stood to her left.

TRINITY SITE

WHERE

THE WORLD'S FIRST

NUCLEAR DEVICE

WAS EXPLODED

JULY 16, 1945

ERECTED 1965

WHITE SANDS MISSILE RANGE

MAJOR GENERAL J. FREDERICK THORLIN

U.S. ARMY COMMANDING

TRINITY SITE HAS BEEN DESIGNATED A NATIONAL
HISTORIC LANDMARK
THIS SITE POSSESSES NATIONAL SIGNIFICANCE IN
COMMEMORATING
THE HISTORY OF THE UNITED STATES OF AMERICA
NATIONAL PARK SERVICE
UNITED STATES DEPARTMENT OF INTERIOR

As I read, a smug laugh began to play inside of me, and I thought, we Americans, so proud of our violence. Out of the corner of my eye, I saw the woman's body shaking. Projecting my nervousness, I wondered if she too were laughing, but then I heard a stifled sob. She moved away and knelt on the ground. Her husband bent over her protectively, murmuring words of comfort. I had planned to photograph the monument. But seeing the weeping woman, I could not. For me, the obelisk became a gravestone and snapping a picture a desecration. Chastened by my own cynicism, I did not judge the picture takers.

Beyond the obelisk was a Fat Man bomb casing, a duplicate of the one tested at Trinity. Walking toward it, I passed a small, low tangle of concrete and metal, all that remains of the hundred-foot tower that held the unexploded plutonium bomb. The bomb, strapped to a semi-truck trailer, seemed awkward and bulky. However, I realized it was, in fact, an extraordinary feat of miniaturization—just eleven feet long, Fat Man razed the city of Nagasaki and World War II ended. Then I noticed people gathering along a far edge of the fence and headed in that direction. I passed uniformed and civilian missile range staff chatting easily with visitors, answering questions and giving directions. When I reached the chain-link and barbed-wire barrier, I saw photographs hanging from it. It was the same series of Trinity images that I had studied at home as a little girl.

Recalling the great medicine wheels of the Plains Indians, I decided to walk the entire boundary of the site. With my first steps I wondered if prayers have the power to heal past suffering. Along a remote edge of the fence, I knelt to examine some small green fragments of Trinitite, and saw next to them, in a perfect square, four rabbit droppings. As the sun rose higher in the cloudless sky, I slowly traveled the circle. When the shimmering Sierra Oscura once again entered my field of vision, I became aware of a rhythmic phrase pushing itself toward the front of my consciousness. Soon my mind was forming words from the Ave Maria: "Holy Mary, mother of God, pray for us sinners now and at the hour of our death."

I had learned that prayer when my mother-in-law died, in order to join the family in saying the Rosary on the eve of her funeral. Mamá had prayed the Rosary every morning—"Dios te salve María . . . " I loved learning this appeal to the Madonna, which had meant so much to my mother-in-law, but my trouble with the Hail Mary was the word *sinners*. I figured the church fathers used the concept of sin to keep the people down and guilty. However, at that moment, in the heat of La Jornada del Muerto, it took on new meaning. It became an earnest supplication, an acknowledgment of how easy it is to be less than we should be, how willing we are to act in ways we know are wrong. And I realized that my own prayer had changed. It was not only a petition for the victims of war but about being human: "Nuestra Señora de la Luz, pray for us sinners, for sinners we surely are."

Leaving ground zero, I saw families posing smilingly for snapshots in front of the obelisk and beside the bomb casing. But there were somber faces as well. Back at the parking lot, I passed the makeshift bookstore and souvenir shops manned by missile range personnel. Friendly soldiers and civilians were selling everything from T-shirts and caps with mushroom clouds to mugs, patches, key chains, thimbles, teaspoons, and other collectibles with the Trinity logo—the black obelisk. The beginnings of the "Trinity Lunch" appeared on long tables. Hot, dusty, and tired, I navigated through the crowd, my deep interest replaced by the pressing need to get away. As I exited a long line of cars was being held at the Stallion Gate—as one left, another was allowed to enter.

The Trinity site is also called "Alamogordo." However, this is a misnomer resulting from the fact that ground zero is situated at the northwest corner of the wartime Alamogordo Bombing and Gunnery Range. The site is actually one hundred miles from the town of Alamogordo. Driving west through the village of San Antonio, it occurred to me that it would more properly have been called "San Antonio." The most famous San Antonio/Saint Anthony is pictured carrying the Christ Child. But the first San Antonio, the original desert father, seems a more appropriate patron. This Egyptian mystic entered the wilderness to wrestle with the devil in his own territory and thus regain Paradise.

The scientists wandered some months in this American desert, where they wrestled with nature's forces, creating an irreversible reality—awesome and terrible beyond anything they had envisioned.

Mosaic

THE PROBLEM OF POWER

Hans Bethe once stated, "You may well ask why people with kind hearts and humanist feelings—why they would go and work on weapons of mass destruction."[1] One way to understand this document is to change a few of Bethe's words: you may well ask what *happens* to people with kind hearts and humanist feelings when they work on weapons of mass destruction—especially when those weapons are used. This is what I am trying to understand, not simply because the question of the atomic scientists is personally important and historically significant, but because citizens of the United States are, in very real ways, heirs to this legacy. And what does this inheritance mean? How does it affect our spirit?

American culture understands the meaning of the bomb primarily in terms of its power. When my father died, some newspapers ran the obituary with the headline "Nuclear Physicist Harry Palevsky" or "Particle Physicist Harry Palevsky." However, his *New York Times* obituary read "Harry Palevsky, 71, an A-bomb Developer." My father was not one of the project's scientific leaders. At the time he was only an electrical engineer with a bachelor's degree, yet he was dubbed an "A-bomb developer." This reflects the power and prestige Americans continue to bestow on the Manhattan Project. I have noticed that, in recent obituaries of project scientists, the closer they were to Oppenheimer and the bomb, the more the scientific prowess that accrues.

What of the other side to the power of the bomb, its effect on its victims? For people "with kind hearts and humanist feelings," can the
bomb's consequences ever be out of sight? Nobel laureate Kenzaburo
Oe posed the question in 1964, quoting a journalist from Hiroshima: "Is
the atomic bomb known better for its immense power or for the human
misery it causes?" Thirty years later Oe wrote of his original work:

> At the time of writing the essays in this book I was sadly lacking in
> the attitude and ability needed to recast Hiroshima in an Asian per-
> spective. In that respect I reflected the prevailing Japanese outlook
> on Hiroshima. In response to criticisms from Korea and the Philip-
> pines, however, I have since revised my view of Hiroshima. I have
> focused more on Japan's wars of aggression against Asian peoples,
> on understanding the atomic bombings of Hiroshima and Nagasaki
> as one result of those wars, and on the special hardships suffered by
> the many Koreans who experienced the atomic bombings.

Yet his initial insight into A-bomb victims' transformation remains
significant.

> I have come to realize that they, one and all, possess unique powers
> of observation and expression concerning what it means to be
> human. I have noticed that they understand in very concrete ways
> such words as courage, hope, sincerity, and even "miserable death."
> The way they use these terms makes them what in Japanese has
> traditionally been called "interpreters of human nature," and
> what today would translate as "moralists."[2]

While the atomic scientists have been characterized by some as being
focused on their technological achievements at the expense of their
morality, Michael Walzer opined, "They were driven by a deep moral
anxiety, not (or not most crucially) by any kind of scientific fascination;
they were certainly not servile technicians. On the other hand, they were
men and women without political power or following, and once their
own work was done, they could not control its use." At the same time
Walzer has rejected the kind of moral calculus Hans Bethe makes in his
argument supporting the decision to use the bomb. Walzer stated, "To

kill 278,966 civilians (the number is made up) in order to avoid the deaths of an unknown but probably larger number of civilians and soldiers is surely a fantastic, godlike, frightening, and horrendous act."[3]

On the other side, Paul Fussell spoke from the soldier's standpoint, in order "to indicate the complex moral situation of knowing one's life has been saved because others' have been most cruelly snuffed out." He has asserted that the ethics of the atomic bomb must also be analyzed from the combat experience. The views of those who would have been killed without its use, he insisted, must be part of any moral debate. And to Walzer's argument against utilitarian calculations, David Hendrickson responded, "It is difficult to avoid the conclusion that the refusal to have considered the consequences of not dropping the bomb would have been, in the circumstances, equally 'fantastic, godlike, frightening, and horrendous.' It is difficult, moreover, to resist the conclusion that dropping the first bomb at least did save many more lives than would have been lost had this decision been refused."[4]

The historian Robert Butow, in an October 1997 letter, encouraged me to look at the larger context of twentieth-century war, asserting that war itself will have to be viewed as a major atrocity before we will ever be able to live in peace. In modern warfare no combatant can walk away with clean hands. It seems to me that implicit in all of these arguments is the problem of power. Modern weapons, in particular weapons of mass destruction—the applications of scientific discovery—possess the might once found only in nature or deity. Yet they present us with dilemmas that the gods of heaven and earth never face.

These issues remain relevant beyond the controversies surrounding the bombings. I believe that we must keep in the front of our consciousness the question of power and how we use it. Herbert York made this point when he spoke of the "special evil" of the atomic bomb. To me, this means that the same feelings of revenge, mistrust, and anger that fuel every war can now have devastating consequences while the powerful remain at a safe distance, never required to face the human suffering they cause. We have much more than our arguments over wartime culpability to consider.

On the fortieth anniversary of the Los Alamos lab, Nobel laureate

I. I. Rabi presented a talk titled "How Well We Meant." He reminisced that in 1943 something larger than nationalist patriotism for the United States had driven the Manhattan Project scientists—their patriotism for Western civilization. "The world seemed about to be engulfed by a fanatic, barbarian culture. And it did look then, as it turned out to be, that only through science and its products could western civilization be saved." Rabi asserted, "Well, we saved it, I say this proudly, and I think, truly." He went on to say that this great new power put the United States in a position "to start on a new road to a new world" in the postwar era. Then the question changed. It was no longer about how to save civilization but how to destroy another culture—how to destroy other human beings. "We have lost sight of the basic tenets of all religions—that a human being is a wonderful thing. We talk as if humans were matter."[5]

Bethe spoke of the "technological imperative," quoting Einstein's statement that since the bombs were used on Hiroshima and Nagasaki, "everything has changed, except human thinking." After a brief period of activity in the direction of international control of atomic weapons and atomic power, the bomb was incorporated into our arsenal as just another weapon. Bethe cautioned that in the face of its unprecedented power, decision makers "should not lightly follow the technological imperative."[6]

Freeman Dyson warned of the seductive attraction technology holds not for the policy maker but for the physicist:

> I have felt it myself, the glitter of nuclear weapons. It is irresistible if you come to them as a scientist. To feel it's there in your hands—to release this energy that fuels the stars, to let it do your bidding. . . . It is something that gives people an illusion of illimitable power and it is, in some ways, responsible for all our troubles, I would say— this, what you might call technical arrogance that overcomes people when they see what they can do with their minds.[7]

My father, during his last face-to-face conversation with his dear cousin, Max Neidorf, said of the atomic bomb, "It was a great achievement, but I am not sure it was a good achievement." The *New York Times* obituary, by emphasizing his work on the Manhattan Project, made my father's connection to the A-bomb's power appear more important than his lifetime of fine scientific accomplishment, his work in Atoms for Peace, Pugwash, the Federation of American Scientists, or the countless ways, in his personal and professional relationships, he tried to be an "interpreter of human nature."

THE BOHR PHENOMENON

Modern science makes strong claims against religious belief, yet I remain intrigued with the concepts of salvation and redemption attached to the bomb. I can read a form of religiosity into Niels Bohr's question to Robert Oppenheimer, "Is it big enough?" Bohr thought that he saw, in physics's discovery and invention, a means by which humankind's warring urges would be curbed when all other efforts down through history had failed. I do not question Bohr's genuine humanitarian impulses, nor do I doubt that he possessed deep insight into the potential political dynamics of atomic weapons. Rather, I wonder about the spiritual dimension of his reckoning—his belief that the bomb had the power to convert us, to show us by its supernatural light that we are all brothers and sisters.

A balance of terror is not what Bohr envisioned. Deterrence was not what he hoped for. At war's end, even before war's end, he sought in the danger of the bomb "a universal agreement . . . the abolition of barriers hitherto considered necessary to protect national interests." He thought that the bomb contained an unprecedented unifying power: "In contrast to other issues where history and traditions may have fostered divergent viewpoints, we are here dealing with a matter of the deepest interest to all nations." And he saw the scientists as the unique agents of this conversion, both as makers of the weapon and as the teachers of its universal lessons: "Let us hope that science, which, through the ages, has stood as a symbol of the progress to be obtained by common human striving, by its latest emphasis on the necessity of concord, may contribute decisively to a harmonious relationship between all nations."[8]

Bohr's student and friend, Victor Weisskopf, summarized Bohr's larger view of the work being done by the Manhattan Project by saying that "in spite of death and destruction, there is a positive future for this world, transformed by scientific knowledge." And his disciple, Robert Oppenheimer, recalled that at Los Alamos Bohr spoke with contempt of Hitler, who hoped to enslave Europe with a few hundred planes and tanks: "[Bohr] said nothing like that would ever happen again; and his own high hope that the outcome would be good, and that in this the role of objectivity, friendliness, cooperation, incarnate in science, would play a helpful part; all this was something that we wished very much to believe."[9]

During my travels and my reading, I have been fascinated by the diversity of ways in which Bohr has been interpreted by scientists and scholars. Lise Meitner was close to the Danish physicist in her belief that the destructive power of the bomb would make the world wars of the century's first half impossible to repeat. Ruth Lewin Sime, her biographer, tells us that the codiscoverer of fission understood that science and technology historically have served war's purpose. Yet, like Bohr, she thought that "the ethical traditions of scientific research and international scientific cooperation offered a model for human betterment and understanding." But whereas Bohr believed it was essential that this new weapon be made known to the world, thereby transforming it, Meitner refused to work on the bomb. Otto Robert Frisch, her nephew, recalled, "Meitner was invited to join the team at work on the development of the nuclear-fission bomb; she refused, and hoped until the very end that the project would prove impossible." Sime explained that Meitner's vision of physics, and her self-conception as a physicist, prevented her from working on the bomb. "Meitner wanted no part of deaths anywhere: she could not commit herself and her physics—the two were not distinct—to a weapon of war."[10]

Edward Teller also identifies deeply and personally with science. When we met he told me that science was his religion. Yet he interprets Bohr and the meaning of nuclear weapons in a way very different from Meitner. In a 1998 essay on science and morality, Teller described his own stance on the development of nuclear weapons as an example of the deeper message of Bohr's love of contradiction.

> Today there is perceived to be a strong contradiction between the results of science and the requirements of morality; for instance, the application of science has led to the development of nuclear weapons, while international morality seems to demand that such results never be applied—and that research leading to them should be stopped. I hold a position radically different from the general point of view, believing that contradiction and uncertainty should be embraced.[11]

During my conversation with Joseph Rotblat, he invoked Bohr using just the argument that Teller rejects. In response to my question about whether there had ever been a chance, with the discovery of fission coming so close to the beginning of World War II, that the bomb would not have been developed in one country or another, Rotblat said that there was a chance to prevent the development of the weapon if Bohr's proposals had been followed.

While I am critical of Bohr's notion that somehow scientific ways of knowing the natural world can be applied to human affairs, I must admit that I remain intrigued, if at times unclear, by what he might have meant by the bomb's complementarity. In answer to the question "What, then, is complementarity?" Bohr's biographer, Abraham Pais, stated, "It is the realization that particle and wave behavior are mutually exclusive, yet that both are necessary for a complete description of all phenomena." Pais went on to explain that without explicit reference to physics, the concept can be formulated as "two aspects of a description that are mutually exclusive yet both necessary for a full understanding of what is to be described." And Richard Rhodes, in his history of the atomic bomb, summarized Bohr's conception of its complementarity thus: "The weapon devised as an instrument of major war would end major war."[12]

Victor Weisskopf wrote that although the expansion of science's knowledge of and power over nature was the source of great difficulty today and in the future, Bohr was never overwhelmed by them. "For [Bohr] every difficulty, every conflict contains its solution. The greater the difficulty, the greater the step to surmount it, the greater is the reward which ensues. . . . In his mind, science not only created problems, it also showed the way to overcome them. Science is, in his mind, one of the most advanced forms of human collaboration. It therefore must lead the way to better human relations."[13]

When I met with the physicist Freeman Dyson, he provided a complementary perspective from which to view Bohr. One of the keys to Bohr's unsuccessful plan for establishing international control of atomic weapons was the timing—that an agreement among the United States, England, and the Soviet Union would be in place before the end of World War II. However, Dyson imagines an agreement not among states but among scientists. And he takes the question farther back in time, before the beginning of the war, to Bohr's initial response to the discovery of fission. It was then, he argues, that the failure occurred. Dyson has written that the moral leadership of Bohr and Einstein, whose stature placed them outside national loyalties, might have achieved a consensus among the world's physicists against nuclear weapons. "All I know is that 1939 was the last chance for physicists to establish an ethical tradition against nuclear weapons, similar to the Hippocratic tradition that stopped the leading biologists from promoting biological weapons. The chance was missed, and from that point on the march of history led inexorably to Hiroshima."[14]

During our conversation at the Institute for Advanced Study in Princeton, I asked Dyson to expand on his argument. Fission was discovered at the end of 1938, and although the war did not start until ten months later, Dyson thinks that the crucial period was spring and summer 1939. Although everyone knew that nuclear weapons might be on the horizon, at that point none of the scientists was under any compulsion to work on them. He explained that although the fear of Heisenberg and the German scientists was not completely unfounded, the choice was to either get together with Heisenberg, the Russians, and everybody else while it was still in the open—coming "to an agreement that this is something you don't want to do—or else everybody goes home and builds bombs." While conceding that neither way was risk-free, Dyson asserted that it would have been better at least to try. But the time interval, he emphasized, was very short. In contrast, by 1941 circumstances were very different.

> The decisions were made, which were, I think, at that time really inevitable because you couldn't communicate. There was no way you could go and ask Heisenberg, "Well, are you going to do this or aren't you?" I mean, Heisenberg was in a black box and the people in America were in a different black box. So there was no chance of coming to any sort of meeting of minds. But '39 was different. Then it was the older generation who could have taken the initiative, particularly Einstein and Bohr. And Heisenberg was younger, but he was *the* great man in Germany and then there was of course [nuclear physicist Igor Vasil'evich] Kurchatov in Russia.

When I asked if he meant that men of such stature were capable of bringing their colleagues together, Dyson emphatically responded,

> Yes, and I think they should have. I would say that they really *failed* us at that point. I mean, I read the transcripts of the meeting at George Washington University, which was the first time that nuclear fission was discussed at an international meeting. It was pathetic. There was Fermi and there was Bohr. The two great men. What were they arguing about? The priorities. Bohr was terribly anxious that Frisch should get credit first, having discovered a way to detect directly the energy released by fission. And Fermi was terribly anxious that [physicist Herbert] Anderson should get credit. And they were just fighting for their priority claims. So there was no discussion of the bigger issues at all. And those two were such great men, but they had feet of clay, I'm sorry to say.

A later meeting with the physicist Michael M. May challenged many of my assumptions, not about the past significance of Niels Bohr's question, but about its present and future meaning. As the codirector of Stanford University's Center for International Security and Arms Control and director emeritus of Lawrence Livermore National Laboratory, May is heir to many of the concerns of the atomic scientists. I had read an article in which he set out a strong and logical argument against both abolition and marginalization of nuclear weapons in the post–cold war era.

I was particularly interested in May's thesis because he invoked Winston Churchill, the man who so thoroughly rejected Bohr's arms control proposals during the war. May argued that the deterrent power of nuclear weapons is essential to global security in the twenty-first century. Reflecting on the dynamics underlying the twentieth century, filled with disastrous wars, May challenged the assumption of "abolitionists," who argue that total nuclear disarmament will lead to greater global safety and stability. And he took issue with "marginalizers" of nuclear weapons, "who want to keep a few around but at the edge of polite policy discussion, unseen and unheard from, except in connection with the occasional rogue and pariah state." Referring to the great risk posed by nuclear weapons, May reasoned,

> If we were in a world of stable states with no rival territorial and other interests that could not be dealt with by empowered and respected international institutions, this risk would indeed be the main matter, and marginalization of nuclear weapons would be in order. But we are not.
>
> Contrary to what many wish, the states of the world are not becoming law-abiding citizens of one world, at least not in essential security matters. . . .
>
> Nuclear weapons are not all that is needed to make war obsolete, but they have no real substitute.

No agreed-upon authority existed during the first half of the twentieth century to prevent the world powers from waging great wars over rival geopolitical claims. And, May asserted, we do not have such an authority today.

> The United Nations Security Council? The International Court of Justice? Maybe someday, but not this century or perhaps the next. When have we or any other major nation submitted to these institu-

tions when our central, or sometimes even our peripheral, interests were involved?

Winston Churchill . . . warned at the beginning of the atomic age that safety could be the sturdy child of terror, and that we should not give up atomic weapons until we were sure and doubly sure that we had something better to take the place of terror in that respect. Look around: we have nothing better to take its place.[15]

It seemed to me that by explicitly summoning Churchill, May expressed a worldview in direct opposition to Bohr's regarding the function and meaning of nuclear weapons. When we met in his Stanford office, I explained that my reading of Bohr's question "Is it big enough?" was a moral one: when the nuclear bomb becomes a reality, if it is powerful enough, it will show us that war is insane, because using such an awful weapon could mean destroying not only our enemy but ourselves, and civilization. A big enough bomb would reveal to humanity the irrationality of continuing to wage war. I told May that it was my understanding that his views differed fundamentally from Bohr's. It seemed to me that for May, peace, security, and the abolition of war had to do not with humanity's moral development but with fear—being so afraid that we would act properly. He responded by posing his own question to me. "Are you quite sure that there is no connection between being scared and moral development?" When I replied that I was not sure, he continued,

> I think there is more of a connection. We'd like to think, you know, that we are suddenly enlightened by grace. And we are. But before we can accept that, it helps to have a little deterrent there. You know, for individual moral actions or immoral actions, and for group actions. It just helps to have a little something that causes you to stop and say, Do I really want to do this?

As he spoke I looked up at his office wall, where hung a poster of a Tibetan mandala, entitled *Kalachakra for World Peace*. Mentally contrasting the mandala and the mushroom cloud, I retorted that nuclear weapons are more than a "little something." May quietly responded, "I think they are just barely big enough."

BEING GOD OR SEEING GOD?

While the language and the logic of natural science have historically been closed to theology and philosophy, sacred images of religious fate and mystical predestination have arisen around the bomb. We are able to call the bomb test Trinity without experiencing any dissonance. But the Christian Trinity is, after all, the mystery of the Godhead. What does this language tell us about Robert Oppenheimer's belief in the transcendent meaning of science's creation?

In an attempt to gain insight into Oppenheimer's spiritual self-concept in relationship to the bomb, I went back to some of his literary references. First I read John Donne's holy sonnet whence he took the name "Trinity."

> Batter my heart, three person'd God; for, you
> As yet but knocke, breathe, shine, and seeke to mend;
> That I may rise, and stand, o'erthrow mee,'and bend
> Your force, to breake, blowe, burn and make me new.
> I, like an usurpt towne, to'another due,
> Labour to'admit you, but Oh, to not end,
> Reason your viceroy in mee, mee should defend,
> But is captiv'd, and proves weake or untrue,
> Yet dearely'I love you, and would be lov'd faine,
> But am betroth'd unto your enemie,
> Divorce mee,'untie, or breake that knot againe,
> Take mee to you, imprison mee, for I
> Except you'enthrall mee, never shall be free,
> Nor ever chast, except you ravish mee.[16]

I was struck by the violence of the religious imagery. In order to be made God's own, the poet asks to be battered. He implores God to do more than "knocke, breathe, shine and seeke to mend," but to expose him to the full, violent, divine force, so that he may be made new. It is only through being ravished that the supplicant will become chaste.

Was Oppenheimer seeing himself, the world, or both, as in need of God's violent cleansing? Did he believe that the bomb was God's way of breaking us so completely that we would be purified and cease to wage war?

In anticipation of the test, Oppenheimer chose the name Trinity. Of his actual experience of it he recalled, "We knew the world would not be the same. A few people laughed, a few people cried. Most people were silent. I remembered the line from the Hindu scripture, the Bhagavad Gita: Vishnu is trying to persuade the prince that he should do his duty, and to impress him takes on his multi-armed form and says: 'Now I am become death, the destroyer of worlds.' I suppose we all thought that, one way or another." Some accounts include Oppenheimer's closing sentence, "There was a great deal of solemn talk that this was the end of the great wars of the century."[17]

But what was Oppenheimer thinking? At first glance, it appears he was arrogantly asserting that the scientists themselves had "become death" by harnessing nature's destructive power. My reading of the Bhagavad Gita leads me to believe that Oppenheimer was sincerely saying something different. As a scholar, he would have known the context of Vishnu's statement. The question is, at that world-changing moment, was the physicist being God or seeing God?

The Bhagavad Gita, or Song of the Lord, is a seven-hundred-verse, self-contained section of the great Sanskrit epic, the *Mahabharata*. The Gita is a dialogue between the prince Arjuna and the god Krishna, avatar of Vishnu, one of the three supreme deities in the Hindu Trimurti. Krishna has incarnated to help Arjuna and his brothers regain their kingdom, as they have been unable to do so peacefully. Because of the obstinacy of the rival king, Lord Krishna's efforts to justly restore the brothers' kingdom also fail. Thus war is the only means remaining. Krishna is serving as Arjuna's charioteer, but the prince does not want to fight.

One scholar explains that the Gita is not a philosophical tract but a religious text "whose purpose is to engender and consolidate certain attitudes in its audience, in much the same way as the 'Lord' of the title, Krishna, attempts in a variety of ways to lead his interlocutor, Arjuna, from perplexity to understanding and correct action." The prince is faced with a dilemma. His duty as a warrior is to fight, but in so doing, he will slaughter his enemies, who are also his relatives. But by refusing to fight, he disrupts the natural and social order.[18]

The Gita opens with the prince asking Krishna to draw his chariot between the two warring armies. As Arjuna gazes over the battlefield on the vast Kurukshetra plain, he expresses the deepest of antiwar sentiments.

> Ah, my Lord! I crave not for victory, nor for kingdom, nor for any
>   pleasure.
> What were a kingdom or happiness or life to me,
> When those for whose sake I desire these things stand here about to
>   sacrifice their property and their lives:
> Teachers, fathers and grandfathers, sons and grandsons, uncles,
>   fathers-in-law, brothers-in-law and other relatives.
> I would not kill them, even for the three worlds; why then for this
>   poor earth? It matters not if I myself am killed. . . .
> We are worthy of a nobler feat than to slaughter our relatives . . .
>   for, my Lord, how can we be happy if we kill our kinsmen?
> Although these men, blinded by greed, see no guilt in destroying
>   their kin, or fighting against their friends, .
> Should not we, whose eyes are open, who consider it to be wrong to
>   annihilate our house, turn away from so great a crime?[19]

It is interesting to note that the Gita is a dialogue more superior, in Oppenheimer's opinion, to those of Plato. In this case the one being questioned is not a human like Socrates but a god who reveals his great power at Arjuna's request. After hearing Krishna's teachings on being and duty, Arjuna asks the god to show himself in his supreme manifestation. The story reaches a climax as Krishna takes on the form of the multiarmed Vishnu, in order, as Oppenheimer explained, to persuade the prince to do his duty. The text describes the wondrous nature of the divine revelation, concluding with,

> If the light of a thousand suns should all at once rise into the sky,
> that might approach the brilliance of that great self.

The son of Pandu [Arjuna] saw the entire universe, in its multi-
plicity, gathered there as one in the body of the god of gods.

Vishnu tells Arjuna,

I am time run on, destroyer of the universe, risen here to annihilate
worlds. Regardless of you, all these warriors, stationed in opposing
ranks, shall cease to exist. Therefore go to it, grasp fame! And having conquered your enemies, enjoy a thriving kingship. They have
already been hewn down by me: . . . simply be the instrument.[20]

And later, addressing the prince as a beloved friend, Krishna confides
that it is through his grace and power that Arjuna has been allowed to
witness what no other mortal has ever seen. It is only then that the warrior, who has refused to fight, takes up his bow and goes into battle.
After eighteen days, Arjuna has vanquished his enemy, but only the five
brothers, a single ally, and Lord Krishna have survived the slaughter.

The exploding atomic bomb must have brought to Oppenheimer's
mind the transfiguration of the charioteer-god Krishna into the
epiphany of the deity Vishnu in all his terrible splendor. Vishnu is the
"destroyer of worlds" and Oppenheimer is the prince. Vishnu has
already killed Arjuna's enemies. The forces at work are larger and more
ancient than human comprehension. It is the prince's duty to carry out
an action that has been fated by God.

Did Oppenheimer see the Trinity test as the Gita made manifest?
Science had brought him face-to-face with the destructive power of the
universe. Did he then believe that he had to do his duty and kill his enemies—enemies who had already been annihilated by the god? Did he
think that this creation of science, and his role in bringing it forth, had
been fated?

It is hard for me to imagine that the horrors of World War II could
have ended in anything but a terrible way. That a shocking new order of
weapon was needed to bring a swift end to a brutal conflict is a strong
argument. Perhaps for that brief moment in history, with the awful
confluence of forces abroad on the earth, science's weapon would seem
to be destiny.

AN ATOMIC SCIENTIST'S APPEAL

On July 25, 1995, within two weeks of the half-century mark of the bombings of Hiroshima and Nagasaki, the "Atomic Scientists' Appeal" was released by the Federation of American Scientists in Hiroshima. It was inspired by a letter from Hans Bethe, the most senior living atomic scientist.

> As the Director of the Theoretical Division at Los Alamos, I participated at the most senior level in the World War II Manhattan Project that produced the first atomic weapons.
>
> Now, at age 88, I am one of the few remaining such senior persons alive. Looking back at the half century since that time, I feel the most intense relief that these weapons have not been used since World War II, mixed with the horror that tens of thousands of such weapons have been built since that time—one hundred times more than any of us at Los Alamos could ever have imagined.
>
> Today we are rightly in an era of disarmament and the dismantlement of nuclear weapons. But in some countries nuclear weapons development still continues. Whether and when the various Nations of the world can agree to stop this is uncertain. But individual scientists can still influence this process by withholding their skills.
>
> Accordingly, I call on all scientists in all countries to cease and desist from work creating, developing, improving and manufacturing further nuclear weapons—and, for that matter, other weapons of potential mass destruction such as chemical and biological weapons.[21]

WHAT SCIENCE IS AND WHAT SCIENCE MAKES

In April 1997 I wrote to Hans Bethe to tell him that I had dreamed about him and science. I found that in addition to the question of the bomb, I was thinking about science's larger meaning. One night I dreamed that Bethe was married to a gypsy woman, who was, for lack of a better word, the spirit of science. After thinking about the dream, I wrote to him:

Often I think of science in technological terms—of the cold machinery, the devices, the accelerators, the weapons that science makes possible—all the

*things that modern science creates and utilizes. However, one day, I thought
of science and appreciated its intent to look more closely into the beauty and
mystery of nature. I had a glimpse of science in a different light, and at that
moment the image of the woman in my dream came to mind. In one view of
science the image exists of the male scientist exerting power and control over
passive female nature. In this view the practice of science is seen as a violation
of the natural world. However, my dream image raised the possibility of an
alternative view. I began to consider another generating impulse of pure science—one born of curiosity and the love of nature. Then the woman becomes
an intriguing symbol of a new way for me to think about the practice of science and its relationship to nature. She embodies the sense of science as the
desire to understand nature, pursued in a rational and imaginative way. . . .*

*Perhaps the dream image embodies the idea of the primary activity of sci-
ence as a romance with nature. Science is then not about the power of (male)
intellect over passive (female) embodied nature. Rather science is a marriage,
the relationship between human intellect and the intelligibility of a dynamic
nature—nature which is both mysterious and knowable and in whose knowing we learn something about ourselves.*

In his May 1997 response, Bethe wrote, "Yes, you have got the essence
of science. It is the desire to understand nature, not the things that science creates." Then he posed the question "How can we make that clear
to people?" I took his question to heart. Why do so many of us, why do
I, the daughter of scientists, hold such negative views of science?

In June 1997 my husband, Joseph, received a call from his sister who
was caring for his ninety-year-old mother. How is it that we recognize
these death calls when they come? We quickly packed our suitcases and
caught an early morning flight to northern California. His mother, Rita,
was dying. We sat vigil in her hospital room as her sister, children,
grandchildren, and great-grandchildren prayed over her and said good-
bye. The first night, we went to the hospital at midnight to relieve
Joseph's sister. It was our good fortune that the second bed in the room
was unoccupied. We both sat for a while, and then Joseph decided to
sleep. At first I tried to read, but I did not want too much light in

Mamá's face, so I put the book aside. As I sat I meditated on Hans Bethe's question and some time later set down my thoughts in a letter to him.

The image of a braid came to my mind. It is as if three conceptual strands are intertwined in my consciousness. The first strand is science, the second is the bomb, and the third is environmental degradation. All three forces have exerted great power during my lifetime. Scientific discovery has resulted in greater technological power, weapons of mass destruction, more control over nature, and unprecedented environmental damage. The application of scientific knowledge has paradoxically resulted in serious harm to and possible ultimate destruction of the natural world that science, at its heart, loves. Your point is that science is not what science makes. I realized that although the three strands have been wound together, I had to separate them in my own mind. I visualized unbraiding the strands, straightening them out, and placing them side by side on the table so that I could look at each more clearly. Then I could begin to understand them as discrete, albeit intimately related, questions. If I separate the strands to understand each better, I must also acknowledge that they are, in fact, intertwined in our consciousness and the sociopolitical world of the late twentieth century. It is crucial that we understand both the separateness of the strands and the ways in which they overlap, influence, and connect to each other.

Two days later Mamá died. Since that time I have thought more about what the braided strands might represent. Does the question go beyond distinguishing between what science is and what science has made? Technology itself has become so tied to the tools of war on a global scale that we assume technoscientific advances must, by definition, result in advances in weaponry. But human beings choose to bring the fruits of scientific discovery into being, and we again choose to use these fruits to bellicose ends. Hans Bethe called on his scientific colleagues to stop work on weapons of mass destruction. But will we as a culture be able, even willing, to at least question the link between our technological prowess and war making?

Even as I write I can hear the arguments about how dangerous the world is now—that better and smarter weapons are needed to keep us safe. But, as Bethe observed, the technological imperative all too often drives the choices—the power is difficult to resist. Traditionally, the decision makers have been the experts while the general public is seen to lack the understanding necessary to make sound judgments. We may not be able to make scientific judgments, but we can make rational judgments about how the consequences will affect our lives. Therefore, while we should not be naive about dangers in the world, we must nonetheless challenge and critically examine powerful assumptions born in the violence of the twentieth century. The question is, will we be able to decouple technological advances from their martial applications?

I have also begun to worry about other strands in the braid. We are in the infancy of understanding our relationship to science and its creations. There is no question in my mind that technology has made my life easier than my forebears'. Science's applications have contributed to my health, leisure, and independence. At the same time it is easy to look back at the discovery of fission, the development of atomic energy, and to judge the scientists as naive in their beliefs about what it meant for humankind's future, or cynical in the manipulation of their discovery's power. Many saw redemption in the atom's promise of universality. But what of future scientific discovery? Do young scientists give any thought to the dilemmas their elders faced? We are already seeing potent applications of the life sciences. Will we go forward without any self-reflection about how a hundred years of unprecedented scientific and technological development have shaped our consciousness and our culture? Will we leave all the decisions to the academic and industrial researchers, as if all the applications of their science were inevitable? And seeing the next avatar in biotechnology, will we make our bodies the laboratory, believing that because biology is the science of life, our lives will be redeemed by what it makes?

LIFE UNDERSTOOD BACKWARD*

I have chosen to keep the memory of my mother and father alive through a discussion of the atomic bomb. This work can be understood

---

* A nod to Kierkegaard.

as a continuation of a dialogue my parents began many years ago. The bomb's meaning was an important element of my father's life as a nuclear physicist. However, the dialogue did not dominate our lives; it took the form of an underground current that would sometimes flow to the surface and cross our paths.

Actually, my most prominent memories of my parents' teaching about the world of science had nothing to do with the bomb. They sought to instill the spirit of internationalism in their five children. In our household science had very much to do with people—my dad's colleagues from around the globe and his journeys abroad. The family sometimes traveled with him. My parents used these visits and trips as opportunities to teach us about other cultures and the importance of communication and tolerance. For them, science was the language that could cross national boundaries and form links between peoples. It was the bridge by which food, art, music, and literature could be shared and learning would take place.

One example of this was the thousand cranes. In the early 1960s a Japanese scientist spent a year working with my father at Brookhaven. As was customary for my father's visiting colleagues, Dr. Wakuta often came for dinner or to spend a weekend afternoon. After he returned to Japan, we received a package containing one thousand origami cranes suspended in strands from a flat, woven basket. My parents explained that each day that Dr. Wakuta had been at Brookhaven, his wife and young daughter had folded three cranes. By the end of the year, they had made one thousand. The Wakuta family sent the cranes to my parents as a gift of thanks. They explained that the crane was a symbol of good fortune and long life. I imagined Dr. Wakuta's wife and daughter folding three cranes each day as a prayer for him, to make sure that he was well and would return home safely. It was not until many years later that I learned that the traditional thousand cranes had become a symbol of the atomic bombs' victims.

At about the same time, my father worked for a summer as a visiting scientist at Los Alamos and the family went along. He was a pioneer in the field of medium-energy physics, and at that time Los Alamos was building the world's largest medium-energy research facility. The half-mile-long linear accelerator is known as LAMPF—Los Alamos Meson Physics Facility—and Dad was the first chairman of the LAMPF visiting scientists' users group.

I do not think my mother had returned to New Mexico since the war. I recall all of us going to the museum at Los Alamos where replicas of Fat Man and Little Boy were displayed in an open-air courtyard formed by the L-shape of the museum. I remember my parents stating very strongly and clearly that they thought it was wrong that the bombs were exhibited as they were. It was one of those strange childhood moments. I knew that they were trying to make a point, but I was not sure exactly what it was—they did not spell it out for us. To me, the replicas meant nothing. They were just funny-looking metal things on stands. I only remember my mother saying that the exhibit was reminiscent of a Japanese garden. Perhaps she was responding to the exhibit's stark simplicity; perhaps there were pebbles surrounding the bombs' platforms, reminding her of a Buddhist rock garden. I cannot recall such physical details, but I vividly recollect my mother purposefully raising her voice to emphasize and make clear to her children that she thought it was "just awful" that the bomb models were displayed in this manner.

I did not have to reconsider my view of my parents' relationship to the bomb until the late 1960s, when I went away to college. A new friend was dating a prominent campus radical, and I occasionally spent time with them and their crowd. One day we were discussing our backgrounds, and I told them about my father's work at Brookhaven lab and, with some pride, my parents' war work on the bomb. My companions were shocked. The radical leader challenged me: Didn't I know how terrible the bomb was? Didn't I understand that my parents had collaborated in one of the worst atrocities of the twentieth century? Furthermore, he asserted, my father's work at a national laboratory was a continuation of the same evil. I do not recall my response. I was shy, and today I imagine that I said nothing.

Another memory is more vivid, perhaps because on this occasion I was not simply the recipient of another person's opinion. This was the day my voice entered the family dialogue about the bomb. It was early on a summer evening at the end of my first year at the university. My father and I were on the patio, engaged, uncharacteristically, in a vehement argument. I was determined to do something that he forbade. I cannot now recall what it was. Failing to change his mind, I became frustrated and angry. But the words I spoke had nothing to do with the dispute at hand. "What gives you the right to decide what is good or bad for me to do? You worked on atomic bombs that killed 150,000 people. And you are always telling me what you know about being a good person. Look at what you have done!"

I succeeded in silencing my father. It was my mother who spoke. Because she was the one with whom I usually fought, and she had the quick temper, I was expecting her anger. She looked straight at me, but I was unable to meet her gaze. I had no explanation. Like my parents, I was shocked by what had been in my heart and had come out of my mouth. Bracing for the worst, I heard her gently ask, "Don't you realize how your father feels? Don't you know what he has been doing? Don't you understand that he is working very hard to make sure it will never happen again?"

I did not understand and therefore could never express the sense of betrayal I felt when my college friends told me what they claimed were the hidden truths about my family. I thought that my parents had deceived me. But I could not say any of this, and my father did not know how to respond to my fury. He just turned and walked inside. I did not get what I wanted, and he never discussed the incident with me.

However, soon after that my father began giving me different kinds of books to read: Yasunari Kawabata's *Snow Country*, Erik Erikson's biography of Mahatma Gandhi, and, later, Yukio Mishima's *Runaway Horses*. I remember sitting at the kitchen table as he explained Erikson's theory of the stages of human development and Gandhi's philosophy of nonviolent resistance. I took Mishima on the ferry ride to Fire Island beach and pondered for the first time the idea of reincarnation. My father talked to me about how important it was for him to learn about a worldview so different from his own. And I recall a telephone call when I returned to school in California. He had been reading Simone Weil. She was a great thinker, he told me, relating something of her philosophy. He soon sent me a volume of her work, which I faithfully read. But

I could never admit to my father that I did not understand what she was saying. Perhaps it is these silences, the words not spoken, the questions not asked, that underlie my study.

FAREWELL

In the late 1970s my father suffered a series of small strokes. He continued to work, but by 1981 his illness forced him to retire. My dad's condition slowly deteriorated, and by summer 1990 his silence had grown deeper. My mother's unexpected death was a grave blow to him. In the year and a half between their deaths, his children, family, and friends did the best they could to bring him comfort. Joseph and I tried to create for my father a home where he could enjoy those things that had always given him pleasure: classical music, art, gardening, good food and wine, and collecting his beloved pottery.

Although he remained an avid reader, I was distressed to find him spending days in front of the television, watching not only news programs and movies but hours of cartoons. I drew strength from a verse in the Old Testament Apocrypha,

O son, help your father in his old age,
and do not grieve him as long as he lives;
even if he is lacking in understanding, show forbearance.

And I was comforted by a passage in Louise Erdrich's novel *Love Medicine*. Her young protagonist, Lipsha Morrissey, describes his grandfather's "second childhood" as a calling, a smoke screen, something he put on himself, so that he could think hard about life. I wanted to believe that my dad was in there, silently pondering the world and our place in it, as he hid behind the irritating laughter of cartoon voices.[22]

At the same time, because of his illness, much of his concern, and my attention, was devoted to his body. He had advanced cardiovascular disease, and the circulation to one of his feet was very poor. My father's pain was chronic, sometimes excruciating. When it was very bad, his concentration, already weakened, broke.

It was in this context that we began work on his memoirs. He approached the project seriously. We sat at a table in my living room, and he spoke slowly and carefully into the microphone. He said that there were certain important things he wanted to record and began by telling me his story about receiving a scholarship to Northwestern University. Whenever he came to a memory of particular significance, he would stop and say, "There is something you should know."

Of course, he told me stories about the Manhattan Project, the Met Lab, Los Alamos, marrying my mother in Santa Fe, and the bomb. It would be a fine ending to this discussion if I were to write that my father and I had a conversation during which the bomb's meaning was resolved and made clear. And it would have been a wonderful end-of-life story for my father if I could say that he had come to terms with the weapon and in so doing had given me a way to do the same. But that it is not what occurred. At the end of one of our conversations about Los Alamos, I asked him what it had been like when the news of Hiroshima arrived. The decision, he said, had been made by an inner circle. There were a lot of mixed feelings. "We wanted to do all we could for the Japanese scientists." He paused and then, with a sob caught in his throat, asked, "Did you know that Nagasaki was the center of Christianity in Japan?" I replied that I did not.

We never spoke of it again. Within a few months of our conversation his health rapidly deteriorated. On September 17, 1990, one day after his seventy-first birthday, my father died. I understood that his question represented his awareness of the psychological, cultural, historical, and spiritual complexity underlying the use of the bomb. Just as he handed me Simone Weil, he left me this question. The enigma became mine to solve.

Nagasaki is the Japanese city with which Westerners have had the longest contact, a trading center whose port was opened in the sixteenth century, the center of Christianity in Japan. Jesuit priests brought the word of the Christian God, the God of Love, the three person'd God, the God of the Golden Rule. The blast of the atomic bomb destroyed nearly two square miles in the heart of the city—70,000 injured, 70,000 dead. Where does it all fit in sacred teachings about the meaning of suffering? What does it mean to have caused such great suffering?

Figures 18–20. Nagasaki, one week after the atomic bombing, August 1945.

On August 6, 1995, the fiftieth anniversary of Hiroshima, I attended a peace retreat at La Casa de María in Santa Barbara, California. On the first night Stella Matsuda performed "Dance of a Thousand Cranes." The following morning we held discussions, and in the afternoon there was a poetry workshop. My late parents were much on my mind. The leader asked each of us to complete the opening phrase, "I want to know . . . " Then we were to read our poems to the group. Weary from the day's heated debates about the bomb, what came to my mind was, "I want to know, Mother, I want to know, Dad, will we meet again?"

As I spoke the words, I wept and wondered if my colleagues would find it strange that a woman in her forties continued to mourn her parents. After the event a gentleman from India approached me and said that he wanted to thank me. He represented an Indian peace group and on returning home planned to write an article saying that he had discovered that Americans love their families. He told me that many Indians believe that we do not honor our parents, and he was glad to learn otherwise.

NOTES

PROLOGUE: BROKEN VESSEL

1. For an in-depth discussion of the role of women scientists in the development of the atomic bomb, see Caroline L. Herzenberg and Ruth H. Howes, "Women of the Manhattan Project," *Technology Review* (November/December 1993): 32–40; Caroline L. Herzenberg and Ruth H. Howes, *Their Day in the Sun: Women of the Manhattan Project* (Philadelphia: Temple University Press, 1999).

2. Spencer R. Weart and Gertrud Weiss Szilard, eds., *Leo Szilard: His Version of the Facts*, vol. 2, *The Collected Works of Leo Szilard* (Cambridge, Mass.: MIT Press, 1978), 149, 181.

3. Alice Kimball Smith, *A Peril and a Hope: The Scientists' Movement in America, 1945–47* (Chicago: University of Chicago Press, 1965), Jeffries Report, 553; Franck Report, 563, 566–567. The signatories to the Franck Report were J. Franck, chairman, D. J. Hughes, J. J. Nickson, E. Rabinowitch, G. T. Seaborg, J. C. Stearns, and L. Szilard.

CHAPTER 1. HANS A. BETHE, TOUGH DOVE

1. Otto Hahn referred to his experimental results as "barium fantasies" in his December 28, 1938, letter to Lise Meitner, quoted in Ruth Lewin Sime, *Lise Meitner: A Life in Physics* (Berkeley: University of California Press, 1996), 239.

2. The historian Mark Walker challenged this thesis about Bothe's mistake. He pointed out that after the war Werner Heisenberg wanted to be recognized as a leader of German physicists and in a 1946 draft article,

asserted that Bothe's mistake had hindered the German project. Walker argued that even though Bothe objected and Heisenberg changed his wording, "this theme, 'if only we had tried graphite . . . ,' continued to circulate within Heisenberg's circle and beyond. Eventually Bothe became a scapegoat—the scientist who had made 'the mistake' that had kept the Germans from achieving a chain reaction." Mark Walker, "Heisenberg, Goudsmit and the German Atomic Bomb," *Physics Today* 43, no. 1 (1990): 52, 55, 56.

I asked Hans Bethe what he thought of the scapegoat argument. In January 1999 he wrote that Walker's article had not changed his opinion.

> Of course, the Germans knew that graphite was a possible moderator for a nuclear reactor. But Bothe made a mistake in measuring its properties. . . . Bothe was justly blamed for apparently showing that graphite was unsuitable, so that the very expensive heavy water had to be used by the Germans. This indeed doomed the German atomic project to failure. But Bothe's mistake was very subtle, how could he have known about boron carbide?

3. For deep analysis of Japan's decision to surrender, see Robert J. C. Butow, *Japan's Decision to Surrender* (Stanford: Stanford University Press, 1954). In his October 1997 correspondence with me, Butow asserted that it is important to develop a sense of the Japanese enemy of that day, as understood by Truman and other American leaders. He did not mean that the atrocities committed by the Japanese in China and Southeast Asia justified the United States' use of the bomb. Rather, his point was that the Americans were influenced, in part, by their conception of the leaders in Tokyo, in estimating the chances of persuading them to surrender, thus avoiding a bloody invasion.

4. Laurens van der Post, *The Prisoner and the Bomb* (New York: William Morrow, 1971). This beautifully written little volume chronicles van der Post's harrowing experiences in a Japanese prisoner-of-war camp and the meaning of the bomb for the captives liberated by it.

5. Quoted in Ronald W. Clark, *Einstein: The Life and Times* (New York: Avon, 1971), 672.

6. Niels Bohr's notes to his July 3, 1944, and March 24, 1945, memoranda to President Roosevelt, May 8, 1945, "Frankfurter-Bohr" file, box 34, J. Robert Oppenheimer Papers, Library of Congress, Washington, D.C.

7. Hans A. Bethe, *The Road from Los Alamos* (New York: AIP Press, 1991), 153.

8. Mary Palevsky Granados, "The Tough Question Will Always Remain: Did We Have to Use the Bomb?" *Los Angeles Times Magazine*, June 25, 1995, 10–11, 28–30.

CHAPTER 2. EDWARD TELLER, HIGH PRIEST OF PHYSICS

1. Edward Teller, *Better a Shield than a Sword: Perspectives on Defense and Technology* (New York: Free Press, 1987), 57.

2. Quoted in Martin J. Sherwin, *A World Destroyed: Hiroshima and the Origins of the Arms Race* (New York: Vintage, 1987), 296.

3. Teller, *Better a Shield*, 58. There is a discrepancy in the dates of this correspondence between Teller and Szilard. Teller explained, "Szilard's letter was dated July 4, 1945, while my reply, dated July 2, was written a number of days after I received his. In addition, Szilard had not bothered to fill in my name in his form letter. The explanation is simply that Szilard was a man of many idiosyncrasies." Teller, *Better a Shield*, 244 n. 7.1. The editors of Szilard's papers ordered the correspondence chronologically and did not describe Teller's July 2 letter as a response to Szilard's petition but rather as part of the general debate about it. See pp. 208–212 in Weart and Szilard, eds., *Leo Szilard*. Regarding the date discrepancy, Szilard's biographer opined, "A more likely explanation is that he [Teller] knew about the petition in advance and drafted his reply on the second [July 2], even before receiving a copy." William Lanouette, *Genius in the Shadows: A Biography of Leo Szilard* (New York: Charles Scribners' Sons, 1992), 527 n. 36.

4. Sherwin, *A World Destroyed*, "Notes of the Interim Committee Meeting, May 31, 1945," 302.

5. Richard G. Hewlett and Oscar E. Anderson, Jr., *A History of the United States Atomic Energy Commission*, vol. 1, *The New World, 1939/1946* (University Park: Pennsylvania State University Press, 1962), 358.

6. Sherwin, *A World Destroyed*, "Science Panel: Recommendations on the Immediate Use of Nuclear Weapons, June 16, 1945," 305.

7. Gar Alperovitz, *The Decision to Use the Atomic Bomb and the Architecture of an American Myth* (New York: Alfred A. Knopf, 1995).

8. Teller, *Better a Shield*, 4–5.

9. Quoted in Herbert F. York, *The Advisors: Oppenheimer, Teller and the Superbomb* (Stanford: Stanford University Press, 1989), 161.

10. Teller, *Better a Shield*, 59.

11. Teller, *Better a Shield*, 59.

12. Teller is referring to Rhodes's analysis of his reasons for working on and role in the development of the H-bomb. See Richard Rhodes, *The Making of the Atomic Bomb* (New York: Simon & Schuster, 1986); and *Dark Sun: The Making of the Hydrogen Bomb* (New York: Simon & Schuster, 1995).

13. Edward Teller, "The Atomic Scientists Have Two Responsibilities," *Bulletin of the Atomic Scientists* 3, no. 12 (1947): 356.

MARTYRS TO HISTORY?

1. Robert J. Maddox, *Weapons for Victory: The Hiroshima Decision Fifty Years Later* (Columbia: University of Missouri Press, 1995).

2. *Notes on Initial Meeting of Target Committee*, April 27, 1945, Los Alamos National Laboratory Archives.

3. Quoted in Butow, *Japan's Decision to Surrender*, 174.

4. Butow, *Japan's Decision to Surrender*, 174.

5. Teller, *Better a Shield*, 4–5.

CHAPTER 3. PHILIP MORRISON, WITNESS
TO ATOMIC HISTORY

1. Philip Morrison, "Recollections of a Nuclear War," *Scientific American*, August 1995, 45.

2. U.S. Senate, Special Committee on Atomic Energy, *Atomic Energy: Hearings on S. Res. 179*, pt. 2, 79th Cong., 1st. sess., December 6, 1945, 233–234, 235, 236.

3. U.S. Senate, Special Committee on Atomic Energy, *Atomic Energy: Hearings on S. Res. 179*, pt. 5, 79th Cong., 2d sess., February 15, 1946, 534.

4. U.S. Senate, *Atomic Energy Hearings*, pt. 2, 241, 243, 244.

5. Morrison, "Recollections of a Nuclear War," 44–45.

6. For a thorough examination of McCarthyism and academia, including Morrison's case, see Ellen W. Schrecker, *No Ivory Tower: McCarthyism and the Universities* (New York: Oxford University Press, 1986).

7. At Morrison's suggestion, I obtained copies of the Target Committee minutes that list participants, but not Philip Morrison. However, they do record, "During the course of the meeting panels were formed from the committee members and others to meet in the afternoon and develop conclusions to items discussed in the agenda." Major J. A. Derry and Dr. N. F. Ramsey to General L. R. Groves, May 12, 1945, *Summary of Target Committee Meetings on 10 and 11 May 1945*, Los Alamos National Laboratory Archives, 1. I guessed that Morrison was one of the unnamed "other" panel members. There is no reference to the discussion he described. In January 1999 I corresponded with Robert S. Norris, author of a forthcoming biography of

Leslie R. Groves, who told me that he has searched the archival record for more documentation of the Target Committee meetings. Although there are some memos to the 20th Air Force and General Curtis LeMay, Norris has concluded, "As to more details about what transpired [at the meetings], I don't think there is anything else."

Perhaps this is an example of historical documents being as important for what they leave out as for what they contain. In a January 1999 letter, Morrison's friend David Hawkins told me that he clearly recalled Morrison talking about the meetings shortly after they occurred. "[Morrison] represented the concerns of many of us! He said that he proposed that a warning be sent to [the] Japanese . . . giving them a chance to evacuate. The officer sitting across from him—name not known, or remembered—spoke vehemently against the proposal, saying something like, 'They'd send up everything they have against us, and I'd be in that plane!' "

8. The publication by John von Neumann to which Morrison refers is "Can We Survive Technology?" *Fortune* 51, no. 6 (1955): 106. Morrison's own blueprint for meeting the challenges of the next century can be found in Philip Morrison and Kosta Tsipis, *Reason Enough to Hope: America and the World of the Twenty-first Century* (Cambridge, Mass.: MIT Press, 1998).

9. Quoted in David Hawkins, "Special Tasks," letter to the editor, *Bulletin of the Atomic Scientists* 50, no. 5 (1994): 60.

PACIFIC MEMORIES I

1. Freeman J. Dyson, *Disturbing the Universe* (New York: Harper & Row, 1979), 77–78, 80.

2. W. H. Auden, "The Bomb and Man's Consciousness," in *Hiroshima plus Twenty*, prepared by the *New York Times* (New York: Delacorte, 1965), 128, 131.

CHAPTER 4. DAVID HAWKINS, CHRONICLER
OF LOS ALAMOS

1. Quoted in Jon Else, *The Day after Trinity: J. Robert Oppenheimer and the Atomic Bomb* (Kent, Ohio: PTV Publications, 1980), transcript of motion picture, 26; see also p. 14.

2. David Hawkins, *Project Y: The Los Alamos Story* (Los Angeles: Tomash, 1983), xii–xiii.

3. Hawkins, *Project Y,* xiii–xiv.

4. Quoted in Alice Kimball Smith and Charles Weiner, eds., *Robert Oppenheimer: Letters and Recollections* (Stanford: Stanford University Press, 1995), 290.

5. Arthur H. Compton, *Atomic Quest* (New York: Oxford University Press, 1956), 127–128.

6. Bethe, *The Road from Los Alamos*, xi. See also Bethe, *The Road from Los Alamos*, 30–33; Hans A. Bethe, "Can Air or Water Be Exploded?" *Bulletin of the Atomic Scientists* 1, no. 7 (1946): 2, 14.

7. Tolman, of Caltech, was a leading science adviser to the Manhattan Project. Conant, a chemist and the president of Harvard, chaired the National Defense Research Committee and was the deputy to Vannevar Bush, director of the Office of Scientific Research and Development. A primary architect of the A-bomb project, Conant, with General Groves, signed the original directive to Oppenheimer regarding the establishment of the Los Alamos laboratory.

8. Morrison, "Recollections of a Nuclear War," 45.

9. David Hawkins, "Should the Scientist Take Part in Politics?" *New York Times Magazine*, June 16, 1946, 13, 44–46.

10. Hawkins, "Special Tasks," 60; J. Robert Oppenheimer, "Three Lectures on Niels Bohr and His Times: Part III: The Atomic Nucleus," unpublished Pegram Lecture, August 1963, box 247, J. Robert Oppenheimer Papers, Library of Congress, Washington, D.C., 9.

11. Niels Bohr to President Roosevelt, July 3, 1944, "Frankfurter-Bohr" file, box 34, J. Robert Oppenheimer Papers, Library of Congress, Washington, D.C., 7.

12. Oppenheimer, "Niels Bohr and His Times," 11; McGeorge Bundy, *Danger and Survival: Choices about the Bomb in the First Fifty Years* (New York: Random House, 1988), 115; Martin J. Sherwin, "Niels Bohr and the First Principles of Arms Control," paper presented at Niels Bohr: Physics and the World, Niels Bohr Centennial Symposium, American Academy of Arts and Sciences, Cambridge, Mass., November 12–14, 1985, 320; Abraham Pais, *Niels Bohr's Times, in Physics, Philosophy, and Polity* (Oxford: Clarendon Press, 1991), 503.

CHAPTER 5. ROBERT R. WILSON, THE PSYCHE
OF A PHYSICIST

1. Jeremy Bernstein, "Physicist: I. I. Rabi—II," *New Yorker*, October 20, 1975, 66.

2. Robert Jungk, *Brighter Than a Thousand Suns: A Personal History of the Atomic Scientists*, translated by James Cleugh (New York: Harcourt Brace Jovanovich, 1958); Robert R. Wilson, review of *Brighter Than a Thousand Suns: A Personal History of the Atomic Scientists*, by Robert Jungk, *Scientific American*, December 1958, 146.

3. Robert R. Wilson, "Niels Bohr and the Young Scientists," *Bulletin of the Atomic Scientists* 41, no. 8 (1985): 25.

4. Oppenheimer quoted in Nuel Pharr Davis, *Lawrence and Oppenheimer* (New York: Simon & Schuster, 1968), 257–258; Truman quoted in Rhodes, *Dark Sun*, 205.

PROFESSOR BETHE AT HOME IN HIS OFFICE

1. Bethe, *The Road from Los Alamos*, 25.

2. Sime, *Lise Meitner*, x.

3. The " Farm Hall Transcripts" can be found in *Operation Epsilon: The Farm Hall Transcripts* (Berkeley: University of California Press, 1993). The original clandestine recordings, along with the bulk of the original German transcripts, no longer exist. The twenty-four weekly Farm Hall reports are English translations of the original German transcripts, made at the time. Therefore, only one true transcript is extant: Werner Heisenberg's lecture of August 14, 1945. See pp. 1, 147.

4. Sime, *Lise Meitner*, 318–319, 322.

5. Sime, *Lise Meitner*, 319, 482 n. 45; Thomas Powers, *Heisenberg's War: The Secret History of the German Bomb* (New York: Alfred A. Knopf, 1993), xi.

6. Bethe, *The Road from Los Alamos*, 16–17.

7. Victor F. Weisskopf, *The Joy of Insight: Passions of a Physicist* (New York: Basic Books, 1991), 165.

CHAPTER 6. JOSEPH ROTBLAT, PUGWASH PIONEER

1. Joseph Rotblat, *Pugwash, the First Ten Years: History of the Conferences on Science and World Affairs* (London: Heinemann, 1967), 77–79. The signatories to the Russell-Einstein Manifesto were Russell, Einstein, Rotblat, Max Born, Percy W. Bridgman, Leopold Infeld, Frédéric Joliot-Curie, Herman J. Muller, Linus Pauling, Cecil F. Powell, and Hideki Yukawa.

2. Joseph Rotblat, "Leaving the Bomb Project," *Bulletin of the Atomic Scientists* 41, no. 8 (1985): 17.

3. After meeting with Rotblat, I looked into this question. Bohr, in urging a postwar agreement to Roosevelt, wrote, "Quite apart from the questions of how soon the weapon will be ready for use and what role it may play in the present war, this situation raises a number of problems which call for most urgent attention." Bohr to Roosevelt, July 3, 1944, 5. Martin Sherwin asserted that Bohr "never opposed the wartime use of the bomb." Sherwin, "Niels Bohr," 321. McGeorge Bundy concluded that Bohr "had no trouble, then or later, with the wartime effort and its wartime purpose; to that effort he made his own contribution." Bundy, *Danger and Survival*, 114.

4. Rotblat, "Leaving the Bomb Project," 18.

5. U.S. Atomic Energy Commission, *In the Matter of J. Robert Oppenheimer: Transcript of Hearing before the Personnel Security Board*, April 12, 1954–May 6, 1954, 173.

6. Rotblat, "Leaving the Bomb Project," 18.

7. Rhodes reported that it was Enrico Fermi, apparently privately, who at a conference in April 1943 initially proposed to Oppenheimer the idea that radioactive fission products from the chain-reacting pile might be used to poison the German food supply. Oppenheimer then carried the idea to General Groves and James Conant. Rhodes, *Making of the Atomic Bomb*, 510–511. On May 25, 1943, Oppenheimer responded by letter to Fermi, "To summarize then, I should recommend delay if that is possible. (In this connection I think that we should not attempt a plan unless we can poison food sufficient to kill a half a million men, since there is no doubt that the actual number affected will, because of non-uniform distribution, be much smaller than this.) If you believe that such delay will be serious, I should recommend discussion with a few well-chosen people." J. Robert Oppenheimer to Enrico Fermi, May 25, 1943, box 33, J. Robert Oppenheimer Papers, Library of Congress, Washington, D.C.

8. Freeman J. Dyson, *From Eros to Gaia* (London: Penguin, 1993), 256; Freeman J. Dyson, *Weapons and Hope* (New York: Harper & Row, 1984), 131.

9. Joseph Rotblat, "Reminiscences on the Fortieth Anniversary of the Russell-Einstein Manifesto," presidential address at the 45th Pugwash Conference on Science and World Affairs, Hiroshima, Japan, July 1995, 1.

10. Joseph Rotblat, "The Nobel Lecture: Remember Your Humanity," Oslo, December 10, 1995.

11. Joseph Rotblat, "The Multifaceted Social Conscience of Scientists," presidential address at the 44th Pugwash Conference on Science and World Affairs, Kolymbari, Crete, June 1994, 188–189.

12. Joseph Rotblat, "Societal Verification," in *A Nuclear-Weapon-Free World: Desirable? Feasible?* edited by Joseph Rotblat, Jack Steinberger, and Bhalchandra Udgaonkar (Boulder, Colo.: Westview Press, 1993), 115.

13. Joseph Rotblat, "Allegiance to Humanity," presidential address at the 46th Pugwash Conference on Science and World Affairs, Lahti, Finland, September 1996, 3–4.

14. Martha C. Nussbaum, "Patriotism and Cosmopolitanism," in Martha C. Nussbaum with Respondents, *For Love of Country: Debating the Limits of Patriotism,* edited by Joshua Cohen (Boston: Beacon Press, 1996), 2–17.

15. Sissela Bok, "From Part to Whole," in Nussbaum with Respondents, *For Love of Country,* 39, 40.

CHAPTER 7. HERBERT F. YORK, INSIDE HISTORY

1. Herbert F. York, *Making Weapons, Talking Peace: A Physicist's Odyssey from Hiroshima to Geneva* (New York: Basic Books, 1987), 8.

2. York, *Making Weapons, Talking Peace,* 20.

3. John Kifner, "Hiroshima: The Controversy That Refuses to Die," *New York Times,* January 31, 1995, A16; Barton J. Bernstein, "The Atomic Bombings Reconsidered," *Foreign Affairs* 74, no. 1 (1995): 149.

4. Quoted in Edward T. Linenthal, "Anatomy of a Controversy," in *History Wars: The Enola Gay and Other Battles for the American Past,* edited by Edward T. Linenthal and Tom Engelhardt (New York: Metropolitan, 1996), 58.

5. York, *Making Weapons, Talking Peace,* 52, 68.

6. Hans A. Bethe, "Comments on the History of the H-Bomb," *Los Alamos Science* 3, no. 3 (1982): 51, 52.

7. Quoted in Ronald W. Clark, *Tizard* (Cambridge, Mass.: MIT Press, 1965), 300–301.

8. Iris Chang, *The Rape of Nanking: The Forgotten Holocaust of World War II* (New York: Basic Books, 1997), 4 (noncombatant dead estimates), 5, 89 (rape estimates); John Rabe, *The Good Man of Nanking: The Diaries of John Rabe,* edited by Erwin Wiekert, translated by John E. Woods (New York: Alfred A. Knopf, 1998), 77.

9. Chang, *The Rape of Nanking,* 15, 6, 5.

10. Asada Sadao, "The Mushroom Cloud and National Psyches: Japanese and American Perceptions of the A-Bomb Decision, 1945–1995," *Journal of American–East Asian Relations* 4, no. 2 (1995): 115.

11. York, *Making Weapons, Talking Peace*, 235.

12. Leo Szilard's November 1961 lecture at Harvard was the starting point of the creation of the Council for a Livable World. See *Toward a Livable World: Leo Szilard and the Crusade for Nuclear Arms Control*, edited by Helen S. Hawkins, G. Allen Greb, and Gertrud Weiss Szilard (Cambridge, Mass.: MIT Press, 1987); Michael Bess, *Realism, Utopia, and the Mushroom Cloud: Four Activist Intellectuals and Their Strategies for Peace, 1945–1989* (Chicago: University of Chicago Press, 1993).

13. Sidney D. Drell and Richard L. Garwin, "Basing the MX Missile: A Better Idea," *Technology Review* (May–June 1981): 20–9; Richard L. Garwin, "Reducing Dependence on Nuclear Weapons: A Second Nuclear Regime," in *Nuclear Weapons and World Politics: Alternatives for the Future*, edited by David C. Gompert, Michael Mandelbaum, Richard L. Garwin, and John H. Barton (New York: McGraw-Hill, 1977); Richard L. Garwin and Hans A. Bethe, "Anti-Ballistic-Missile Systems," *Scientific American*, March 1968, 21–31.

14. Sidney D. Drell et al., "Nuclear Testing: Summary and Conclusions," JSR–95–320 (McLean, Va.: JASON, The MITRE Corporation, 1995).

EPILOGUE: MOSAIC

1. Quoted in Else, *Day after Trinity*, 1.

2. Kenzaburo Oe, *Hiroshima Notes*, translated by David L. Swain and Toshi Yonezawa (New York: Grove, 1996), 67, 9, 78.

3. Michael Walzer, *Just and Unjust Wars: A Moral Argument with Historical Illustrations*, 2d ed. (New York: Basic Books, 1992), 263, 262.

4. Paul Fussell, *Thank God for the Atom Bomb, and Other Essays* (New York: Summit, 1988), 44; David C. Hendrickson, "In Defense of Realism: A Commentary on *Just and Unjust Wars*," *Ethics and International Affairs* 11 (1997): 25–26.

5. I. I. Rabi, "How Well We Meant," in *New Directions in Physics: The Los Alamos 40th Anniversary Volume*, edited by Nicholas Metropolis, Donald M. Kerr, and Gian-Carlo Rota (Boston: Academic Press, 1987), 263, 264.

6. Hans A. Bethe, "The Technological Imperative," *Bulletin of the Atomic Scientists* 41, no. 8 (1985): 34, 36.

7. Quoted in Else, *Day after Trinity*, 30.

8. Niels Bohr, "A Challenge to Civilization," *Science*, October 12, 1945, 363, 364.

9. Victor F. Weisskopf, "Niels Bohr, the Quantum and the World," in

Niels Bohr: A Centenary Volume, edited by A. P. French and P. J. Kennedy (Cambridge, Mass.: Harvard University Press, 1985), 27; Oppenheimer, "Niels Bohr and His Times," 9.

10. Sime, *Lise Meitner,* 375, 306; Otto Robert Frisch, "Lise Meitner," in *Dictionary of Scientific Biography,* edited by Charles C. Gillespie (New York: Charles Scribner's Sons, 1974), 9:263.

11. Edward Teller, "Science and Morality," *Science,* May 22, 1998, 1200.

12. Pais, *Niels Bohr,* 23, 24; Rhodes, *Making of the Atomic Bomb,* 532.

13. Victor F. Weisskopf, "Niels Bohr and International Scientific Collaboration," in *Niels Bohr: His Life and Work as Seen by His Friends and Colleagues,* edited by S. Rozenthal (New York: John Wiley & Sons, 1967), 265.

14. Freeman J. Dyson, "Non-Use and Non-Violence," paper presented at the Oak Ridge Symposium on Non-Use of Nuclear Weapons, Oak Ridge, Tenn., May 1995, 7.

15. Michael M. May, "Fearsome Security: The Role of Nuclear Weapons," *Brookings Review* 13, no. 3 (1995): 25, 26, 27.

16. "Holy Sonnet XIV," in Helen Gardner, ed., *John Donne: The Divine Poems,* 2d ed. (Oxford: Clarendon Press, 1978), 11.

17. Quoted in Else, *Day after Trinity,* 30; quoted in Len Giovannitti and Fred Freed, *The Decision to Drop the Bomb: A Political History* (New York: Coward-McCann, 1965), 197.

18. W. J. Johnson, *The Bhagavad Gita: A New Translation* (Oxford: Oxford University Press, 1994), vii; Narasimhan's prose translation was most helpful in providing an overview of the story underlying the Gita. Chakravarthi V. Narasimhan, *The Mahābhārata,* rev. ed. (New York: Columbia University Press, 1998).

19. Shri Purohit Swami, *The Bhagavad Gita: The Gospel of the Lord Shri Krishna* (New York: Alfred A. Knopf, 1977), 9.

20. Johnson, *Bhagavad Gita,* 49–50, 51. This translation differs from Oppenheimer's "Now I am become death, the destroyer of worlds."

21. Hans A. Bethe, Open Letter, in "Atomic Scientists Appeal to Colleagues: Stop Work on Further Nuclear Weapons," *F.A.S. Public Interest Report* 48, no. 5 (1995): 8. The appeal was endorsed by the FAS Council and had the individual endorsements of atomic scientists Marvin L. Goldberger, Jerome Karle, Glenn T. Seaborg, Philip Morrison, Victor Weisskopf, Robert R. Wilson, and Herbert York and an endorsement by Richard L. Garwin.

22. Eccles. 3:12–13, Revised Standard Edition; Louise Erdrich, *Love Medicine* (New York: Holt, Rinehart & Winston, 1984), 190–191.

SELECTED BIBLIOGRAPHY

Interviews were conducted by the author during the period 1995–1998. Transcripts, audiotapes, and notes are retained by the author.

Akira Iriye. "Historical Scholarship and Public Memory." *Journal of American–East Asian Relations* 4, no. 2 (1995): 89–93.

Alperovitz, Gar. *The Decision to Use the Atomic Bomb and the Architecture of an American Myth*. New York: Alfred A. Knopf, 1995.

Asada Sadao. "The Mushroom Cloud and National Psyches: Japanese and American Perceptions of the A-Bomb Decision, 1945–1995." *Journal of American–East Asian Relations* 4, no. 2 (1995): 95–116.

Atomic Bomb Fiftieth Anniversary (Special Issue). *Journal of American History* 82, no. 3 (1995).

Auden, W. H. "The Bomb and Man's Consciousness." In *Hiroshima plus Twenty*, prepared by the *New York Times*, 126–132. New York: Delacorte, 1965.

Auden, W. H., and Christopher Isherwood. *The Ascent of F6: A Tragedy in Two Acts*. London: Faber & Faber, 1936.

Badash, Lawrence, Joseph O. Hirschfelder, and Herbert P. Broida, eds. *Reminiscences of Los Alamos: 1943–1945*. Dordrecht: D. Reidel, 1980.

Bainbridge, Kenneth T. "A Foul and Awesome Display." *Bulletin of the Atomic Scientists* 31, no. 5 (1975): 40–46.

Baker, Paul R., ed. *The Atomic Bomb: The Great Decision*. New York: Holt, Rinehart & Winston, 1976.

Barnett, Teresa. "Analyzing Oral Texts, or, How Does an Oral History Mean?" *Oral History Review* 18, no. 2 (1990): 109–113.

Barzun, Jacques. "History, Popular and Unpopular." In *The Interpretation of History*, edited by Joseph R. Strayer, 29–57. New York: Peter Smith, 1950.

Batchelder, Robert C. *The Irreversible Decision: 1939–1950*. Boston: Houghton Mifflin, 1962.

Bateson, Mary Catherine. *Composing a Life*. New York: Penguin, 1990.

———. *Peripheral Visions*. New York: HarperCollins, 1994.

Behar, Ruth, and Deborah A. Gordon, eds. *Women Writing Culture*. Berkeley: University of California Press, 1995.

Berlin, Isaiah. *The Crooked Timber of Humanity: Chapters in the History of Ideas*. Edited by Henry Hardy. New York: Alfred A. Knopf, 1991.

Bernstein, Barton J. "The Atomic Bombings Reconsidered." *Foreign Affairs* 74, no. 1 (1995): 135–152.

———. Introduction to *Toward a Livable World: Leo Szilard and the Crusade for Nuclear Arms Control*, edited by Helen S. Hawkins, G. Allen Greb, and Gertrud Weiss Szilard, xvii–lxxiv. Cambridge, Mass.: MIT Press, 1987.

———. "Misconceived Patriotism." *Bulletin of the Atomic Scientists* 51, no. 3 (1995): 4.

Bernstein, Jeremy. *Hans Bethe, Prophet of Energy*. New York: Basic Books, 1980.

———. "Physicist: I. I. Rabi—I." *New Yorker*, October 13, 1975, 47–110.

———. "Physicist: I. I. Rabi—II." *New Yorker*, October 20, 1975, 47–102.

———. "What Did Heisenberg Tell Bohr about the Bomb?" *Scientific American*, May 1995, 92–97.

Bertaux, Daniel, ed. *Biography and Society: The Life History Approach in the Social Sciences*. Beverly Hills, Calif.: Sage, 1981.

Bess, Michael. *Realism, Utopia, and the Mushroom Cloud: Four Activist Intellectuals and Their Strategies for Peace, 1945–1989*. Chicago: University of Chicago Press, 1993.

Bethe, Hans A. "The American Hydrogen Bomb." In *The Atomic Age: Scientists in National and World Affairs*, edited by Eugene Rabinowitch, 144–155. New York: Basic Books, 1963.

———. "Can Air or Water Be Exploded?" *Bulletin of the Atomic Scientists* 1, no. 7 (1946): 2, 14.

———. "Comments on the History of the H-Bomb." *Los Alamos Science* 3, no. 3 (1982): 43–53.

———. Open Letter in "Atomic Scientists Appeal to Colleagues: Stop Work on Further Nuclear Weapons." *F.A.S. Public Interest Report* 48, no. 5 (1995): 8.

———. *The Road from Los Alamos.* New York: AIP Press, 1991.

———. "The Technological Imperative." *Bulletin of the Atomic Scientists* 41, no. 8 (1985): 34–36.

Bethe, Hans A., Kurt Gottfried, and Roald Z. Sagdeev. "Did Niels Bohr Share Nuclear Secrets?" *Scientific American*, May 1995, 85–90.

Bohr, Niels. "A Challenge to Civilization." *Science*, October 12, 1945, 363–364.

Bok, Sissela. "From Part to Whole." In Martha C. Nussbaum with Respondents, *For Love of Country: Debating the Limits of Patriotism*, edited by Joshua Cohen, 38–44. Boston: Beacon Press, 1996.

Boorse, Henry A., Lloyd Motz, and Jefferson Hane Weaver. *The Atomic Scientists: A Biographical History.* New York: John Wiley & Sons, 1989.

Bridgman, Percy. W. "Scientists and Social Responsibility." *Bulletin of the Atomic Scientists* 4, no. 3 (1948): 69–72.

Broad, William J. "Rewriting the History of the H-Bomb." *Science*, November 19, 1982, 769–772.

Broyard, Anatole. *Intoxicated by My Illness, and Other Writings on Life and Death.* Compiled and edited by Alexandra Broyard. New York: Fawcett Columbine, 1992.

Bruner, Edward. M. "The Ethnographic Self and the Personal Self." Introduction to *Anthropology and Literature*, edited by Paul Benson, 1–26. Urbana: University of Illinois Press, 1993.

Bundy, McGeorge. *Danger and Survival: Choices about the Bomb in the First Fifty Years.* New York: Random House, 1988.

Bundy, McGeorge, William J. Crowe, and Sidney D. Drell. *Reducing Nuclear Danger: The Road Away from the Brink.* New York: Council on Foreign Relations, 1993.

Butow, Robert J. C. *Japan's Decision to Surrender.* Stanford: Stanford University Press, 1954.

Cassidy, David C. *Uncertainty: The Life and Science of Werner Heisenberg.* New York: W. H. Freeman, 1992.

Chang, Iris. *The Rape of Nanking: The Forgotten Holocaust of World War II.* New York: Basic Books, 1997.

Church, Peggy Pond. *The House at Otowi Bridge.* Albuquerque: University of New Mexico Press, 1959.

Clark, Ronald W. *Einstein: The Life and Times.* New York: Avon, 1971.

———. *Tizard.* Cambridge, Mass.: MIT Press, 1965.

Coles, Robert. *The Call of Stories: Teaching and the Moral Imagination.* Boston: Houghton Mifflin, 1989.

Compton, Arthur H. *Atomic Quest*. New York: Oxford University Press, 1956.

Compton, Arthur H., and Farrington Daniels. "A Poll of Scientists at Chicago: July, 1945." *Bulletin of the Atomic Scientists* 4, no. 2 (1948): 44, 63.

Conway, Jill Ker. *The Road from Coorain*. New York: Vintage Books, 1990.

———. *True North*. New York: Vintage Books, 1995.

Corson, Dale R. "Research to Protect, Restore and Manage the Environment." Washington, D.C.: Academy Press, 1993.

Crawford, Elisabeth, Ruth Lewin Sime, and Mark Walker. "A Nobel Tale of Postwar Injustice." *Physics Today* 50, no. 9 (1997): 26–32.

Davis, Nuel Pharr. *Lawrence and Oppenheimer*. New York: Simon & Schuster, 1968.

Denzin, Norman K. *Interpretive Biography*. Newbury Park, Calif.: Sage, 1989.

———. *Interpretive Ethnography*. Thousand Oaks, Calif.: Sage, 1997.

Devereux, George. *From Anxiety to Method in the Behavioral Sciences*. Paris: Mouton, 1967.

Drell, Sidney D. *In the Shadow of the Bomb: Physics and Arms Control*. New York: AIP Press, 1993.

———. "Remarks at Trinity Conference." Paper presented at the Trinity Conference, National Academy of Sciences, Washington, D.C., July 16, 1995.

Drell, Sidney D., and Richard L. Garwin. "Basing the MX Missile: A Better Idea." *Technology Review* (May–June 1981): 20–29.

Drell, Sidney D., et al. "Nuclear Testing: Summary and Conclusions." JSR–95–320. McLean, Va.: JASON, The MITRE Corporation, 1995.

Dyson, Freeman J. *Disturbing the Universe*. New York: Harper & Row, 1979.

———. *From Eros to Gaia*. London: Penguin, 1992.

———. "Non-Use and Non-Violence." Paper presented at the Oak Ridge Symposium on Non-Use of Nuclear Weapons, Oak Ridge, Tenn., May 1995.

———. *Weapons and Hope*. New York: Harper & Row, 1984.

Edel, Leon. *Writing Lives: Principia Biographica*. New York: W. W. Norton, 1984.

Else, Jon. *The Day after Trinity: J. Robert Oppenheimer and the Atomic Bomb*. Kent, Ohio: PTV Publications, 1980. Transcript of motion picture.

Erdrich, Louise. *Love Medicine*. New York: Holt, Rinehart & Winston, 1984.

Feis, Herbert. *The Atomic Bomb and the End of World War II*. Rev. ed. Princeton: Princeton University Press, 1966.

Fermi, Rachel, and Esther Samra. *Picturing the Bomb: Photographs from the Secret World of the Manhattan Project*. New York: Harry N. Abrams, 1995.

Fitch, Val. "The View from the Bottom." *Bulletin of the Atomic Scientists* 31, no. 2 (1975): 43–46.

Frank, Richard B. *Guadalcanal: The Definitive Account of the Landmark Battle*. New York: Penguin, 1990.

Friedman, Hideko Tamura. "Hiroshima Memories." *Bulletin of the Atomic Scientists* 51, no. 3 (1995): 16–22.

Frisch, Michael. *A Shared Authority: Essays on the Craft and Meaning of Oral and Public History*. Albany: State University of New York Press, 1990.

Frisch, Otto Robert. "Lise Meitner." In *Dictionary of Scientific Biography*, edited by Charles C. Gillespie, 260–263. New York: Charles Scribner's Sons, 1974.

Fussell, Paul. *Thank God for the Atom Bomb, and Other Essays*. New York: Summit, 1988.

Gaddis, John Lewis. "On Moral Equivalency and Cold War History." *Ethics and International Affairs* 10 (1996): 131–148.

Gardner, Helen, ed. *John Donne: The Divine Poems*. 2d ed. Oxford: Clarendon Press, 1978.

Garwin, Richard L. "The Maintenance of Nuclear Weapon Stockpiles without Nuclear Explosion Testing." Paper presented at the 24th Pugwash Workshop on Nuclear Forces, Nuclear Forces in Europe, London, September 1995.

———. "Reducing Dependence on Nuclear Weapons: A Second Nuclear Regime." In *Nuclear Weapons and World Politics: Alternatives for the Future*, edited by David C. Gompert, Michael Mandelbaum, Richard L. Garwin, and John H. Barton, 83–139. New York: McGraw-Hill, 1977.

Garwin, Richard L., and Hans A. Bethe. "Anti-Ballistic-Missile Systems." *Scientific American*, March 1968, 21–31.

Giovannitti, Len, and Fred Freed. *The Decision to Drop the Bomb: A Political History*. New York: Coward-McCann, 1965.

Gleick, James. *Genius: The Life and Science of Richard Feynman*. New York: Pantheon, 1992.

Goldberg, Stanley. "Smithsonian Suffers Legionnaires' Disease." *Bulletin of the Atomic Scientists* 51, no. 3 (1995): 28–33.

Gottfried, Kurt, and Bruce G. Blair, eds. *Crisis Stability and Nuclear War*. New York: Oxford University Press, 1988.

Grele, Ronald J., et al. *Envelopes of Sound: The Art of Oral History*. Rev. ed. Chicago: Precedent Publishing, 1985.

Haldane, J. B. S. *Daedalus or Science and the Future: A Paper Read to the Heretics, Cambridge, February 4, 1923.* London: Kegan, Paul, 1924.

Hallas, James H. *The Devil's Anvil: The Assault on Peleliu.* Westport, Conn.: Praeger, 1994.

Hallion, Richard P., and Herman S. Wolk. "Air and Space Museum Guilty as Charged." Letter to the Editor. *Bulletin of the Atomic Scientists* 51, no. 4 (1995): 75.

Hamabata, Matthews Masayuki. *Crested Kimono: Power and Love in the Japanese Business Family.* Ithaca: Cornell University Press, 1990.

Hawkins, David. *Project Y: The Los Alamos Story.* Los Angeles: Tomash, 1983.

———. "Should the Scientist Take Part in Politics?" *New York Times Magazine*, June 16, 1946, 13, 44–46.

———. "Special Tasks." Letter to the Editor. *Bulletin of the Atomic Scientists* 50, no. 5 (1994): 60.

Hawkins, Francis P. Lothrop. *Journey with Children: The Autobiography of a Teacher.* Niwot: University Press of Colorado, 1997.

Hawkins, Helen S., G. Allen Greb, and Gertrud Weiss Szilard, eds. *Toward a Livable World: Leo Szilard and the Crusade for Nuclear Arms Control.* Vol. 3, *The Collected Works of Leo Szilard.* Cambridge, Mass.: MIT Press, 1987.

Heilbrun, Carolyn G. *Writing a Woman's Life.* New York: Ballantine, 1989.

Hendrickson, David C. "In Defense of Realism: A Commentary on *Just and Unjust Wars.*" *Ethics and International Affairs* 11 (1997): 19–54.

Hersey, John. *Hiroshima.* New York: Alfred A. Knopf, 1946.

Hershberg, James G. *James B. Conant: Harvard to Hiroshima and the Making of the Nuclear Age.* New York: Alfred A. Knopf, 1993.

Herzenberg, Caroline L., and Ruth H. Howes. *Their Day in the Sun: Women of the Manhattan Project.* Philadelphia: Temple University Press, 1999.

———. "Women of the Manhattan Project." *Technology Review* (November–December 1993): 32–40.

Hewlett, Richard G., and Oscar E. Anderson, Jr. *A History of the United States Atomic Energy Commission.* Vol. 1, *The New World, 1939/1946.* University Park: Pennsylvania State University Press, 1962.

Hexter, Jack H. *The History Primer.* New York: Basic Books, 1971.

Hodes, Elizabeth. "Precedents for Social Responsibility among Scientists: The American Association of Scientific Workers and the Federation of American Scientists, 1938–1948." Ph.D. dissertation, University of California, Santa Barbara, 1982.

Holloway, David. *Stalin and the Bomb: The Soviet Union and Atomic Energy, 1939–1956.* New Haven: Yale University Press, 1994.

Johnson, W. J. *The Bhagavad Gita: A New Translation*. Oxford: Oxford University Press, 1994.

Jungk, Robert. *Brighter Than a Thousand Suns: A Personal History of the Atomic Scientists*. Translated by James Cleugh. New York: Harcourt Brace Jovanovich, 1958.

Kawabata, Yasunari. *Snow Country*. Translated by Edward G. Seidensticker. New York: Alfred A. Knopf, 1956.

Kevles, Daniel J. *The Physicists: The History of a Scientific Community in Modern America*. New York: Alfred A. Knopf, 1978.

Kifner, John. "Hiroshima: The Controversy That Refuses to Die." *New York Times*, January 13, 1995, A16.

Koestler, Arthur. *The Sleepwalkers: A History of Man's Changing Vision of the Universe*. New York: Macmillan, 1959.

Kohn, Richard H. "History and the Culture Wars: The Case of the Smithsonian Institution's *Enola Gay* Exhibition." *Journal of American History* 82, no. 3 (1995): 1036–1063.

Kondo, Dorinne K. "Dissolution and Reconstitution of Self: Implications for Anthropological Epistemology." *Cultural Anthropology* 1, no. 1 (1986): 74–88.

Lanouette, William. *Genius in the Shadows: A Biography of Leo Szilard*. New York: Charles Scribner's Sons, 1992.

Laue, Max von. "The Wartime Activities of German Scientists." *Bulletin of the Atomic Scientists* 4, no. 4 (1948): 103.

Laurence, William L. "The Scientists: Their Views 20 Years Later." In *Hiroshima plus Twenty*, prepared by the *New York Times*, 114–125. New York: Delacorte, 1965.

L'Engle, Madeleine. *The Summer of the Great-grandmother*. New York: Farrar, Straus & Giroux, 1974.

Lewis, Richard S., and Jane Wilson, eds. *Alamogordo plus Twenty-five Years: The Impact of Atomic Energy on Science, Technology and World Politics*. New York: Viking, 1971.

Libby, Leona Marshall. *The Uranium People*. New York: Crane Russak, 1979.

Lifton, Robert Jay, and Greg Mitchell. *Hiroshima in America: A Half Century of Denial*. New York: G. P. Putnam's Sons, 1995.

Linenthal, Edward T. "Anatomy of a Controversy." In *History Wars: The Enola Gay and Other Battles for the American Past*, edited by Edward T. Linenthal and Tom Engelhardt, 9–62. New York: Metropolitan, 1996.

Linenthal, Edward T., and Tom Engelhardt, eds. *History Wars: The Enola Gay and Other Battles for the American Past*. New York: Metropolitan, 1996.

Lonergan, Bernard J. F. *Insight: A Study of Human Understanding*. New York: Philosophical Library, 1957.

Maddox, Robert J. *Weapons for Victory: The Hiroshima Decision Fifty Years Later*. Columbia: University of Missouri Press, 1995.

Mascia-Lees, Frances E., Patricia Sharpe, and Colleen Ballerino Cohen. "The Postmodernist Turn in Anthropology: Cautions from a Feminist Perspective." In *Anthropology and Literature*, edited by Paul Benson, 225–248. Urbana: University of Illinois Press, 1993.

May, Michael M. "Fearsome Security: The Role of Nuclear Weapons." *Brookings Review* 13, no. 3 (1995): 24–27.

May, Rollo. *The Courage to Create*. New York: Bantam, 1975.

McDaniel, Boyce. "A Physicist at Los Alamos." *Bulletin of the Atomic Scientists* 30, no. 10 (1974): 39–43.

McMahan, Eva M. *Elite Oral History Discourse: A Study in Cooperation and Coherence*. Tuscaloosa: University of Alabama Press, 1989.

McMillan, Priscilla Johnson. Review of *Edward Teller: Giant of the Golden Age of Physics*, by Stanley A. Blumberg and Louis G. Panos. *Scientific American*, May 1990, 130–134.

Merton, Thomas. *The Hidden Ground of Love: Letters on Religious Experience and Social Concerns*. Edited by William H. Shannon. New York: Harcourt Brace Jovanovich, 1985.

———. *The Seven Storey Mountain*. New York: Harcourt, Brace, 1948.

———. *Zen and the Birds of Appetite*. New York: New Directions, 1968.

Mills, C. Wright. *The Sociological Imagination*. New York: Grove Press, 1961.

Mishima, Yukio. *Runaway Horses*. Translated by Michael Gallagher. New York: Alfred A. Knopf, 1973.

Morison, Samuel Eliot. *History of United States Naval Operations in World War II*. Vol. 5, *The Struggle for Guadalcanal, August 1942–February 1943*. New York: Little, Brown, 1959.

Morrison, Philip. "*Alsos:* The Story of German Science." *Bulletin of the Atomic Scientists* 3, no. 12 (1947): 354, 365.

———. *Nothing Is Too Wonderful to Be True*. Woodbury, N.Y.: AIP Press, 1995.

———. "Recollections of a Nuclear War." *Scientific American*, August 1995, 42–46.

———. "A Reply to Dr. von Laue." *Bulletin of the Atomic Scientists* 4, no. 4 (1948): 104.

Morrison, Philip, and Kosta Tsipis. *Reason Enough to Hope: America and the World of the Twenty-first Century*. Cambridge, Mass.: MIT Press, 1998.

Narasimhan, Chakravarthi V. *The Mahābāhrata*. Rev. ed. New York: Columbia University Press, 1998.

Neumann, John von. "Can We Survive Technology?" *Fortune* 51, no. 6 (1955): 106–108, 151–152.

Norris, Robert S. *Racing to the Finish, General Leslie R. Groves, Builder of the Atomic Bomb*. South Royalton, Vt.: Steerforth Press, forthcoming.

Nussbaum, Martha C. "Patriotism and Cosmopolitanism." In Martha C. Nussbaum with Respondents, *For Love of Country: Debating the Limits of Patriotism*, edited by Joshua Cohen, 2–17. Boston: Beacon Press, 1996.

Nussbaum, Martha C., with Respondents. *For Love of Country: Debating the Limits of Patriotism*. Edited by Joshua Cohen. Boston: Beacon Press, 1996.

Oe, Kenzaburo. "The Day the Emperor Spoke in a Human Voice." Translated by John Nathan. *New York Times Magazine*, May 7, 1995, 103–105.

———. *Hiroshima Notes*. Translated by David L. Swain and Toshi Yonezawa. New York: Grove, 1996.

Operation Epsilon: The Farm Hall Transcripts. Berkeley: University of California Press, 1993.

Oppenheimer, J. Robert. "The International Control of Atomic Energy." *Bulletin of the Atomic Scientists* 1, no. 12 (1946): 1–5.

———. Papers. Library of Congress, Manuscript Division, Washington, D.C.

———. "Physics in the Contemporary World." *Bulletin of the Atomic Scientists* 4, no. 3 (1948): 65–68, 85–86.

Pais, Abraham. *Niels Bohr's Times, in Physics, Philosophy, and Polity*. Oxford: Clarendon Press, 1991.

Palevsky Granados, Mary. "The Tough Question Will Always Remain: Did We Have to Use the Bomb?" *Los Angeles Times Magazine*, June 25, 1995, 10–11, 28–30.

Panofsky, W. K. H. *Particles and Policy*. Woodbury, N.Y.: AIP Press, 1994.

Peierls, Rudolf. *Bird of Passage: Recollections of a Physicist*. Princeton: Princeton University Press, 1985.

Polkinghorne, Donald E. *Narrative Knowing and the Human Sciences*. Albany: State University of New York Press, 1988.

Potok, Chaim. *The Book of Lights*. New York: Alfred A. Knopf, 1981.

Powers, Thomas. *Heisenberg's War: The Secret History of the German Bomb*. New York: Alfred A. Knopf, 1993.

———. "Was It Right?" *Atlantic Monthly*, July 1995, 20–23.

Rabe, John. *The Good Man of Nanking: The Diaries of John Rabe.* Edited by Erwin Wickert, translated by John E. Woods. New York: Alfred A. Knopf, 1998.

Rabi, I. I. "How Well We Meant." In *New Directions in Physics: The Los Alamos 40th Anniversary Volume,* edited by Nicholas Metropolis, Donald M. Kerr, and Gian-Carlo Rota, 257–265. Boston: Academic Press, 1987.

Rabinowitch, Eugene. *The Dawn of a New Age: Reflections on Science and Human Affairs.* Chicago: University of Chicago Press, 1963.

Rabinowitch, Eugene, ed. *The Atomic Age: Scientists in National and World Affairs.* New York: Basic Books, 1963.

Rhodes, Richard. *Dark Sun: The Making of the Hydrogen Bomb.* New York: Simon & Schuster, 1995.

———. *The Making of the Atomic Bomb.* New York: Simon & Schuster, 1986.

Rotblat, Joseph. "Allegiance to Humanity." Presidential address at the 46th Pugwash Conference on Science and World Affairs, Lahti, Finland, September 1996.

———."Leaving the Bomb Project." *Bulletin of the Atomic Scientists* 41, no. 8 (1985): 16–19.

———. "The Multifaceted Social Conscience of Scientists." Presidential address at the 44th Pugwash Conference on Science and World Affairs, Kolymbari, Crete, June 1994.

———. "The Nobel Lecture: Remember Your Humanity." Oslo, December 10, 1995.

———. *Pugwash, the First Ten Years: History of the Conferences on Science and World Affairs.* London: Heinemann, 1967.

———. "Reminiscences on the Fortieth Anniversary of the Russell-Einstein Manifesto." Presidential address at the 45th Pugwash Conference on Science and World Affairs, Hiroshima, July 1995.

———. "Societal Verification." In *A Nuclear-Weapon-Free World: Desirable? Feasible?* edited by Joseph Rotblat, Jack Steinberger, and Bhalchandra Udgaonkar, 103–118. Boulder, Colo.: Westview Press, 1993.

———. "Time to Rethink the Idea of World Government." Presidential address at the 42d Pugwash Conference on Science and World Affairs, Berlin, September 1992.

Sachs, Robert G. "Historical Background of the CP-1 Experiment." Paper presented at the Nuclear Chain Reaction—Forty Years Later Conference, University of Chicago, 1984.

Sadako Peace Day, August 6, 1996. *Waging Peace Bulletin* 6, no. 2 (1996): 11.

Sarton, May. *I Knew a Phoenix: Sketches for an Autobiography*. New York: W. W. Norton, 1959.

———. *Plant Dreaming Deep*. New York: W. W. Norton, 1968.

Schrecker, Ellen W. *No Ivory Tower: McCarthyism and the Universities*. New York: Oxford University Press, 1986.

Seaborg, Glenn T. *The Plutonium Story: The Journals of Professor Glenn T. Seaborg, 1939–1946*. Columbus, Ohio: Batelle Press, 1994.

Sherwin, Martin J. "Niels Bohr and the First Principles of Arms Control." Paper presented at Niels Bohr: Physics and the World, Niels Bohr Centennial Symposium, American Academy of Arts and Sciences, Cambridge, Mass., November 12–14, 1985.

———. *A World Destroyed: Hiroshima and the Origins of the Arms Race*. With new introduction. New York: Vintage, 1987. Originally published as *A World Destroyed: The Atomic Bomb and the Grand Alliance*. New York: Alfred A. Knopf, 1975.

Shils, Edward. "Why the Failure?" In *The Atomic Age: Scientists in National and World Affairs*, edited by Eugene Rabinowitch, 76–91. New York: Basic Books, 1963.

Sime, Ruth Lewin. *Lise Meitner: A Life in Physics*. Berkeley: University of California Press, 1996.

Simpson, John A. "Arthur Holly Compton." In *Remembering the University of Chicago: Teachers, Scientists and Scholars*, edited by Edward Shils, 69–84. Chicago: University of Chicago Press, 1991.

———. "The Scientists as Public Educators." *Bulletin of the Atomic Scientists* 3, no. 9 (1947): 243–246.

———."Scientists in Washington: They Did Something about the Bomb." *University of Chicago Magazine* 39 (1946): 3–8.

Smith, Alice Kimball. *A Peril and a Hope: The Scientists' Movement in America: 1945–47*. Chicago: University of Chicago Press, 1965.

Smith, Alice Kimball, and Charles Weiner, eds. *Robert Oppenheimer: Letters and Recollections*. Stanford: Stanford University Press, 1995.

Sodei, Rinjiro. "Hiroshima/Nagasaki as History and Politics." *Journal of American History* 82, no. 3 (1995): 1119–1123.

Spencer, Metta. " 'Political' Scientists." *Bulletin of the Atomic Scientists* 51, no. 4 (1995): 62–68.

Stimson, Henry L. "The Decision to Use the Atomic Bomb." *Harper's*, February 1947, 97–107.

Stimson, Henry L., and McGeorge Bundy. *On Active Service in Peace and War*. New York: Harper & Brothers, 1948.

Stoff, Michael B., Jonathan F. Fanton, and R. Hal Williams, eds. *The Manhattan Project: A Documentary Introduction to the Atomic Age*. Philadelphia: Temple University Press, 1991.

Strickland, Donald A. *Scientists in Politics: The Atomic Scientists Movement, 1945–46*. Lafayette, Ind.: Purdue University Studies, 1968.

Swami, Shri Purohit. *The Bhagavad Gita: The Gospel of the Lord Shri Krishna*. New York: Alfred A. Knopf, 1977.

Szilard, Leo. *The Voice of the Dolphins and Other Stories*. New York: Simon & Schuster, 1961.

Teller, Edward. "The Atomic Scientists Have Two Responsibilities." *Bulletin of the Atomic Scientists* 3, no. 12 (1947): 355–356.

———. *Better a Shield Than a Sword: Perspectives on Defense and Technology*. New York: Free Press, 1987.

———. *The Reluctant Revolutionary*. Columbia: University of Missouri Press, 1964.

———."Science and Morality." *Science*, May 22, 1998, 1200–1201.

———. "Scientists in War and Peace." *Bulletin of the Atomic Scientists* 1, no. 6 (1946): 10–11.

Teller, Edward, with Allen Brown. *The Legacy of Hiroshima*. New York: Doubleday, 1962.

Thompson, Paul. *The Voice of the Past: Oral History*. 2d ed. Oxford: Oxford University Press, 1988.

Truman, Harry S. *Memoirs by Harry S. Truman*. Vol. 1, *Year of Decisions*. New York: Doubleday, 1955.

Tsuzuki, Masao. "Experimental Studies on the Biological Action of Hard Roentgen Rays." *American Journal of Roentgenology* 16, no. 2 (1926): 134–150.

United States Atomic Energy Commission. *In the Matter of J. Robert Oppenheimer: Transcript of Hearing before the Personnel Security Board*. April 12, 1954–May 6, 1954. Washington, D.C.: GPO, 1954.

United States Senate, Special Committee on Atomic Energy. *Atomic Energy: Hearings on S. Res. 179*. 79th Cong., 1st. sess., pt. 2, December 6, 1945. Washington, D.C.: GPO, 1945.

———. *Atomic Energy: Hearings on S. Res. 179*. 79th Cong., 2d. sess., pt. 5, February 15, 1946. Washington, D.C.: GPO, 1946.

van der Post, Laurens. *The Prisoner and the Bomb*. New York: William Morrow, 1971.

Walker, J. Samuel. *Prompt and Utter Destruction: Truman and the Use of Atomic*

Bombs against Japan. Chapel Hill: University of North Carolina Press, 1997.

Walker, Mark. "Heisenberg, Goudsmit and the German Atomic Bomb." *Physics Today* 43, no. 1 (1990): 52–60.

Walzer, Michael. *Just and Unjust Wars: A Moral Argument with Historical Illustrations*. 2d ed. New York: Basic Books, 1992.

Weart, Spencer R., and Gertrud Weiss Szilard, eds. *Leo Szilard: His Version of the Facts*. Vol. 2, *The Collected Works of Leo Szilard*. Cambridge, Mass.: MIT Press, 1978.

Weintraub, Stanley. *The Last Great Victory: The End of World War II, July/August 1945*. New York: Dutton, 1995.

Weisskopf, Victor F. *The Joy of Insight: Passions of a Physicist*. New York: Basic Books, 1991.

———. "Looking Back on Los Alamos." *Bulletin of the Atomic Scientists* 41, no. 8 (1985): 20–22.

———. "Niels Bohr and International Scientific Collaboration." In *Niels Bohr: His Life and Work as Seen by His Friends and Colleagues*, edited by S. Rozenthal, 261–265. New York: John Wiley & Sons, 1967.

———. "Niels Bohr, the Quantum and the World." In *Niels Bohr: A Centenary Volume*, edited by A. P. French and P. J. Kennedy, 19–29. Cambridge, Mass.: Harvard University Press, 1985.

———. *The Privilege of Being a Physicist*. New York: W. H. Freeman, 1989.

White, Lynn. "Science and the Sense of Self: The Medieval Background of the Modern Confrontation." In *Limits of Scientific Inquiry*, edited by Gerald Holton and Robert S. Morrison, 47–59. New York: W. W. Norton, 1979.

Wilber, Ken. *Grace and Grit: Spirituality and Healing in the Life and Death of Treya Killam Wilber*. Boston: Shambhala, 1991.

Wilson, Jane, ed. *All in Our Time: The Reminiscences of Twelve Nuclear Pioneers*. Chicago: Educational Foundation for Nuclear Science, 1975.

Wilson, Jane, and Charlotte Serber, eds. *Standing By and Making Do: Women of Wartime Los Alamos*. Los Alamos, N. Mex.: Los Alamos Historical Society, 1988.

Wilson, Robert R. "Hiroshima: The Scientists' Social and Political Reaction." *Proceedings of the American Philosophical Society* 140, no. 3 (1996): 350–357.

———. "Niels Bohr and the Young Scientists." *Bulletin of the Atomic Scientists* 41, no. 8 (1985): 23–26.

————. "A Recruit for Los Alamos." *Bulletin of the Atomic Scientists* 31, no. 3 (1975): 41–47.

————. Review of *Brighter Than a Thousand Suns: A Personal History of the Atomic Scientists*, by Robert Jungk. *Scientific American*, December 1958, 145–149.

York, Herbert F. *The Advisors: Oppenheimer, Teller and the Superbomb*. Stanford: Stanford University Press, 1989.

————. *Arms and the Physicist*. Woodbury, N.Y.: AIP Press, 1995.

————. *Making Weapons, Talking Peace: A Physicist's Odyssey from Hiroshima to Geneva*. New York: Basic Books, 1987.

————. *Race to Oblivion: A Participant's View of the Arms Race*. New York: Simon & Schuster, 1970.

Zuckerman, Solly. *Nuclear Illusion and Reality*. New York: Viking, 1982.

SOURCES OF ILLUSTRATIONS

CHAPTER 2

Page 43. Courtesy of the Los Alamos Historical Museum Archives (P1989–13-1–283).

CHAPTER 3

Page 75. Courtesy of Los Alamos National Laboratory.

CHAPTER 4

Page 99. Courtesy of David Hawkins.

CHAPTER 5

Page 128. Photo by Don Cooksey. Courtesy of Jane Wilson.

CHAPTER 6

Page 165. Author's collection. Page 177. Courtesy of Joseph Rotblat.

THE OLD COUNTRY

Page 187. Courtesy of Helen Futterman.

CHAPTER 7

Page 201. Courtesy of Jon Brenneis.

EPILOGUE

Page 247. Courtesy of the Los Alamos Historical Museum Archives (M3703e/80.483, M3703f/80.483, M3703g/80.483).

INDEX

nuclear arms control, 188, 193, 211; on nuclear weapons, abolition of, 193, 211–12; on nuclear weapons arsenals, 193, 211–12; Oak Ridge and, 190; on Pearl Harbor, 197; Pugwash and, 197; Rotblat and, 197; Teller and, 200, 205–6; Vietnam War and, 210; on war, 200, 207, 210; Weisskopf and, 189, 190

York, Sybil, 188, 192

Yukawa, Hideki, 255n1 (Rotblat)

Designer:	Barbara Jellow
Compositor:	Binghamton Valley Composition
Text:	Cycles
Display:	Confidential, Gill Sans, Cycles
Printer:	Edwards Brothers
Binder:	Edwards Brothers